A Born Writer

A BORN WRITER

Juanita Harrison
and Her Beautiful World

CATHRYN HALVERSON

University of Massachusetts Press
AMHERST AND BOSTON

Copyright © 2025 by University of Massachusetts Press
All rights reserved

ISBN 978-1-62534-899-9 (paper); 900-2 (hardcover)

Designed by Deste Relyea
Set in Adobe Jenson Pro and Corporative
Printed and bound by Books International, Inc.

Cover design by adam b. bohannon
Cover photo by unknown artist. *Portrait of Juanita Harrison, author of "My Great, Wide, Beautiful World,"* c. 1936. Courtesy of Photographs and Prints Division, Schomburg Center for Research in Black Culture, The New York Public Library.

Library of Congress Cataloging-in-Publication Data
A catalog record for this book is available from the Library of Congress.

British Library Cataloguing-in-Publication Data
A catalog record for this book is available from the British Library.

The authorized representative in the EU for product safety and compliance is
Mare-Nostrum Group. Email: gpsr@mare-nostrum.co.uk
Physical address: Mare-Nostrum Group B.V., Mauritskade 21D, 1091 GC Amsterdam, The Netherlands

TO SÖDERTÖRN UNIVERSITY ENGLISH DEPARTMENT,
1996–2024

Contents

Illustrations ix
Preface xi
Acknowledgments xiii

INTRODUCTION
"My True Life Story" 1

PART I
Before *My Great, Wide, Beautiful World*: American Beginnings 17

CHAPTER 1
Mississippi Ambition 19

CHAPTER 2
Vagabond Housemaid 39

CHAPTER 3
Los Angeles Friendship and Patronage 61

PART II
During *My Great, Wide, Beautiful World*: World Travel and Authorship 89

CHAPTER 4
Writing and Traveling for Freedom 91

CHAPTER 5
Work and Play in France   115

CHAPTER 6
Imperial Geographies   141

PART III
After *My Great, Wide, Beautiful World*: Celebrity and Beyond   161

CHAPTER 7
Hawaii Interludes   163

CHAPTER 8
Authorship and Fame   185

CHAPTER 9
Pan-Hispanic World   210

CHAPTER 10
Last Years   228

EPILOGUE
"A True Literature of American Life"   238

Notes   243
Index   267

Illustrations

Figure 1. George W. Dickinson at the Los Angeles Country Club.   67
Figure 2. Alice Foster Cunningham at Virginia Road School.   83
Figure 3. Mildred Morris in *Peter Pan*.   126
Figure 4. *My Great, Wide, Beautiful World*, original cover.   191
Figure 5. Harrison and Ellery Sedgwick in Honolulu.   197
Figure 6. Harrison with two young Tadas.   200
Figure 7. Postcard from Honolulu.   204
Figure 8. Brazil visa.   221

Preface

I have lived with Juanita Harrison for a long time, since 1995 when I came across an excerpt from her travelogue in Margo Culley's anthology of American women's life narratives, *A Day at a Time*. I considered including her in my dissertation, but my committee member Patsy Yaeger advised me to "save it for the book." I did write about Harrison in the book it became, and in the next two books as well. Her story kept calling me, and I kept finding new ways to think about her and new material to assess.

In the meantime my life continued on. I cannot match my subject for itinerancy, but by academic standards I have had a roving career. Since that first encounter with Harrison I have lived and worked in ten states and seven countries, and traveled commensurately. Often I found she was ahead of me—checking out the Prado, eating on Copenhagen street corners, touring a tea estate outside Nuwara Eliya, staying at the YWCA in Kobe, Japan, so close to my ten-year home in that city that I would pass it on my regular run to the same mountain trails she once hiked.

These points of connection notwithstanding, the gulf between the experience of a Black working-class woman born in Mississippi in the late nineteenth century and that of a white middle-class academic born several generations later cannot be overstated. This gulf accounts for why, at the 2022 meeting of the Biographers International Organization, a member of its board asked me why I should be the one to write this book. She continued, "By the looks of you, you are not Black. People must always ask you this." In fact prior to her they hadn't, at least not directly, but the question is often implicit, and one that I have asked myself. There is much I do not know and cannot speak to, both as a white woman and as a scholar located outside African American studies. Recognizing that other readers have greater insight into Harrison's lived experience, I try to tread carefully. I am sure I do not always succeed. Yet

Harrison herself was inclusive, not exclusive, tolerant of difference and receptive to other perspectives. Crossing all manner of literal and figurative borders, she defied fixed identity categories and celebrated her collaborations and exchanges with people of all nationalities, ethnicities, and creeds. I see this book as another such collaboration, of which I hope she would approve. The first biography is only a beginning, and I look forward to other scholars building on this study to create their own narratives of Harrison's life.

Acknowledgments

I got the idea for the book while a research fellow in the American Studies Department at the University of Wyoming, which gave me shelter after my tenured position at the University of Copenhagen in Denmark vanished in the wake of federal budget cuts. During this scholarly emergency, as I raced (relatively speaking) to complete my previous monograph, *Faraway Women and the "Atlantic Monthly."* I kept trying to wedge more ideas about Harrison into an already overcrowded chapter. This led to a lightbulb moment in my borrowed office—that once again I should "save it for the book," this time, a full biography.

That early fellowship in Wyoming was later bookended by my 2021–22 tenure as Fulbright Distinguished Chair of American Studies at Uppsala University's Swedish Institute for North American Studies. Despite my having the luxury of a full year to devote to the book, the generosity I received in Sweden made me recollect the anxiety Harrison expressed while lingering in the same country: "There are no end to my Friends. I have no time to do any writing." Along with the Fulbright Commission and its staff, I wish to thank Emma Cleary, Helena Wahlström Henriksson, Adam Hjorthén, Christer Larsson, and above all my host, Dag Blanck, for all their help. The book was further supported in Sweden by a generous RJ Sabbatical grant from Riksbankens Jubileumsfond.

Nearing the project's last stage, I could finally take up my fellowship at the Huntington Library to research Harrison's years in southern California, long deferred during the pandemic but coming just when I needed it. Uppsala University's English Department supported research in France. The Fulbright Commission funded a lecture at the Obama Institute in Mainz, Germany, kindly hosted by Alfred Hornung, and a crucial research trip to Honolulu. At the University of Hawaii at Mānoa, John Rosa gave me good leads, while Craig Howe urgently directed me to the county courthouse and made 1930s Honolulu come alive during a long, invested conversation. I wish also to thank Pat Lau, who led me to Harrison's grave at Valley of the Temples Memorial Park.

I appreciate audience responses to presentations at meetings of the American Studies Association, Society for the Study of American Women Writers, the Western Literature Association, and the International AutoBiography Association—most recently, thoughtful counsel from Angela Hooks at the latter's meeting in Reykjavik. Editors and peer reviewers have given me useful feedback on iterations of this work. Former *Journal X* editor Jay Watson merits a note of appreciation, over a quarter of a century after the fact, for his validating response to my first submission about Harrison in 1997: "I love this essay." Incisive reports from anonymous readers at the University of Massachusetts Press have proved invaluable for restructuring, scaffolding, and filling in the gaps. Sharon Brinkman has made my prose more exact. Throughout the process, Brian Halley has been an encouraging and dependable guide—my second book under his watch.

I am grateful as well to the supportive, tight-knit English Department at my second to last job, at Södertörn University in Stockholm, with special thanks to Liz Kella for finding solutions to tricky points in the manuscript. However, in June 2024 we were abruptly informed that by the end of the year the English Department would be shut down, leaving half of us without jobs. I finished this book despite, not because of, the senior management of Södertörn University, whose cruelty and shortsightedness have been breathtaking. This book is dedicated to that bygone department.

Finally, a look back to really starting this project at Minot State University in Minot, North Dakota. There I had a supportive chair, Robert Kibler, and research funding from Vice-President Laurie Geller. I am especially grateful to the Gordon B. Olson Library and its staff, a haven in a cold windy town. It was at a table there that I remember picking up a pen to write the first lines of the introduction, daunted at the thought of the years of work ahead. Still, I feel sad to be done with this book.

INTRODUCTION

"My True Life Story"

Juanita Harrison's travel memoir, *My Great, Wide, Beautiful World*, includes a succinct account of the author's visit with a group of African women villagers. They were in not Africa but Paris, temporarily residing at the Jardin d'Acclimatation on the city's western outskirts. A former zoo, the park had shifted from exhibiting flora and fauna gathered from France's colonial empire to displaying living human beings. Harrison recounts,

> I climbed over the fence and got in the native village where the Plate mouthed women are. a slip is cut just wide enough in the lip to fit around the rim of a wooden plate they can hardly talk with it only the women have the plates there is about ten of them and they took a fancy to me. I think they saw I had some of their blood I couldnt fool them. the yongest wife was during the Cooking as I hung around the Camp fire she offered me some it was good and I would have accepted to save the price of my supper But the spit run out of her mouth on this plate and often dropped into the pot. When I left I climbed over the fence again so it didn't cost me anything.[1]

She expresses a commingled sense of kinship to and difference from these women, connection and remove. Suggesting that their recognition of her African "blood" leads them to "fancy" her, she invites readers to consider whether shared heritage contributed to her own attraction, for why she "hung around the Camp fire." At the same time, her eager participation in this colonialist spectacle puts her Western privilege on display, even as the unpolished prose in which she represents it signals an underprivileged history of her own. Her refusal of the Africans' food underscores their cultural distance, and her hightailing it over the fence that encloses them casts in high relief her liberty as a solo American traveler. On the eve of her departure from Paris, she considered returning to the complex to give the women a winter coat she thought they could use—but decided it was too much trouble. We will revisit this scene, a touchstone in the study, in pages to come.

Incidents such as this one suggest the complexity of Harrison's multilayered subjectivity—woman, worker, tourist, writer, migrant, expatriate, minority, Westerner, American, African American—and the rich texture of a lifetime devoted to travel, leisure, and cross-cultural encounter.[2] This book chronicles that life, along with the social and economic conditions that shaped it. Following Harrison from her birth in Columbus, Mississippi, in 1887 to her death in Honolulu, Hawaii, in 1967, along the way the study makes stops in New York, Havana, Los Angeles, Paris, Cairo, Mumbai, Kobe, Buenos Aires, and many other cities and nations as she circles the globe.

Harrison left Mississippi as a young woman to begin crisscrossing North America, taking a series of domestic jobs and short holidays in the South, Midwest, East, and West as well as in Cuba, Canada, and Hawaii. Her world travels commenced at age thirty-nine, when she set sail for England. She worked and traveled in Europe (especially Spain and France), the Middle East, and Asia for the next eight years before returning to American territory, Hawaii.

My Great, Wide, Beautiful World is a compilation of the letters Harrison sent to friends and past employers over the course of that extended journey—"voluminous batches" of "lengthy epistles," to quote one of its contemporary reviews.[3] Putting her ravenous geographical appetite on display, the text charts both her travels and the short-term jobs as maid, cook, and nurse she took to fund them. A two-part excerpt was published in 1935 in the illustrious *Atlantic Monthly*, once the nation's leading literary magazine. The bound volume that followed was an official bestseller, making Harrison the most commercially successful African American author of the mid-twentieth century—not Richard Wright, who is usually credited with the accomplishment.[4] Edited to appear like a travel journal but with Harrison's original, nonstandard orthography preserved, *My Great, Wide, Beautiful World* updates an established African American tradition of freedom narrative, a singular instance of a memoir by a Black working-class woman that reached mass audiences across the nation and beyond. Harrison settled in Honolulu just before the book was published, and she went on to spend another decade working and traveling in South America before returning to southern California and Hawaii, where she spent the last years of her life.

Juanita Harrison has a unique place in African American history and literature. She was a typical Southern Black woman of her time with respect to the education she received and the work she did. Yet her mobility, which swept her to authorship and celebrity, was exceptional. Her urge to roam was surely a reaction to coming of age in the American South, a region that brutally curbed the movements of its African American residents, in quite spatial as well as social and economic ways. She was among the tens of thousands of Black women and men who left the South in search of greater opportunity, agents of the transformative exodus that became known as the Great Migration. She parted ways with them, however, in that she consistently chose, over the course of more than fifty years, to prioritize her wanderlust. Her relentless travel was at once an expression of self-determination and a symptom of trauma, as she responded with restlessness to oppression and personal loss.

"Freedom is my main point," she declared (315). In the context of African American travel writing, Michael Ra-Shon Hall has defined freedom as "a state of being physically and socially unrestricted." For Harrison, we should add to this formulation the adverbs "economically," "emotionally," and "subjectively."[5] Within the borders of the United States, staying on the move helped her attain a measure of the freedom she yearned for. Ranging outside those borders magnified it.

In acting thus she replicated the "unauthorized liberatory initiatives" of nineteenth-century Black travelers.[6] On leaving their home country, Elizabeth Stordeur Pryor states, they "made the profound discovery that the U.S. style of racism and white supremacy was not natural or innate to white people." No longer called racial epithets on the streets, no longer denied access to public spaces, they learned that "freedom could be heard, seen, inhaled, and touched."[7] These discoveries continued well into the twentieth century. Jessie Redmon Fauset, for example, characterized her stay in Venice as "the first time since I was seventeen that I have been—comparatively—free from fetters."[8] Like these wayfarers, Harrison, too, reveled in new liberties abroad.

I have culled the title of this biography from a statement in which Harrison, quoting her editor Mildred Morris, revives a collocation: "She said that I was a born Writer. I thought how far I was from what I was born" (FL2). Harrison's metaphor about "how far" she had come encourages us to view her literary activity as closely entwined with her mobility with respect to the physical, legal, socioeconomic, and, not least, emotional distance she had traveled from her early life in Mississippi. Note that the statement invokes not the "where" but the "what" of her origins. Her writing enabled her to travel even as her travel inspired her to write and, eventually, to publish. This interlaced practice led her to occupy a range of subject positions and enter a plethora of social relationships both expedient and gratifying.

Colonial histories and ongoing colonial structures in Asia, the Middle East, North Africa, and the Pacific supported her freedom quest. Even as Harrison was subjected, as an African American, to debilitating constraints in her own country, she was granted untold liberties as a Western traveler overseas. That is, she personally benefited from political systems founded on the exploitation of people of color and the "erasure of Black spatial thought," to quote Pat Noxolo.[9] Yet her subject position, and the perspective and behaviors it induced, distinguished her from white fellow citizens. Harrison's identity as a Black working-class woman charges with alternative valence both the travel venture she pursued and the narrative she wrote about it. *My Great, Wide, Beautiful World* shares in the "cultural work" that Gary Totten identifies as characteristic of African American travel writing in the Jim Crow era, that of challenging the bonds placed on their subjects' mobility. The genre has its roots, he states, in a "rich but troubled narrative of black movement," including "the forced travel of the slave trade and the displacement of African Americans within the United States."[10]

Gathering biography, literary criticism, and history, this study poses and seeks to answer a set of questions about Harrison. How could this disenfranchised laborer, a woman who came of age in the late nineteenth century under a crushing racial regime, travel the world and publish a book? What drove her mobility, how could she sustain it, and what kind of subjective claims did she make as a result? Chimene Jackson, author of an essay on Black women's journal writing that features Harrison, pungently encapsulates such an interrogation in questions of her own: "When did she become comfortable with her written voice? . . . When did she develop the cosmopolitan mindset or get fearless in her black skin?"[11]

I cannot account for all the key people, events, and locations in Harrison's life. Thus *A Born Writer* does not look like a biography as usually conceived, disappointing the generic expectations conjured by the term: definitive accounts of forebears, childhood influences, contributions to public life, sunset years. But what is the alternative? To restrict life studies to subjects with enough worldly achievement or access to power to generate copious records across decades? It is largely due to her race and class that Harrison's early history in the United States is so sparsely documented, and her marginal socioeconomic status spurred the inveterate transiency that further contributes to the dearth of evidence. To let her story go untold because the record is incomplete is to perpetuate the cycle. Or to cast this condition more positively, the difficulty impels us to create fresh scholarly genres. This book offers one approach to writing a biography of a subject whose life events are only episodically in view.

I will note that I do not want this biography to be sheer celebration. Everyone who writes about Harrison, in both her day and ours, lauds her achievement: the immense distance she traveled from her origins; the magnitude of her literary success; her shrewdness, wit, and zest; her sticking it to the man and many men; her talent for seizing pleasure wherever she went. Yet of course, as my research has progressed, I have come to recognize Juanita Harrison as a complex human being, with the attendant shortcomings. She was vain. She used people. She lied. She could voice a casual antisemitism. While she was keen to meet strangers, to pass an afternoon or a day in their company, she resisted more enduring relationships. She hurried to the aid of those in need, but she was uninterested in supporting larger social causes. Discussing such tendencies is not censure, but rather acknowledgement of her full humanity. Only to marvel is to do her a disservice.

My Great, Wide, Beautiful World was advertised as a response to the "clamorous demand for More" that the *Atlantic Monthly* excerpts had unleashed the previous year.[12] Running through nine printings within ten months and reviewed in newspapers and magazines nationwide, it drew an enormous and remarkably diverse readership, sparking the enthusiasm of Black working-class library patrons, white women's book clubs, Japanese American journalists, Harlem Renaissance luminaries, and many others. Korey Garibaldi identifies it as the most commercially successful book by an African American author of its time, the first big marketplace success. It was restored to greater visibility in 1996 when reissued in Henry Louis Gates and Jennifer Burton's major G. K. Hall series, "African-American Women Writers 1910–1940." *My Great, Wide, Beautiful World* is known among scholars who study travel writing, life narrative, Black internationalism, and African American women's literature and history.

Nevertheless, Harrison has attracted surprisingly little critical attention. While she is periodically rediscovered by readers who, just like their 1930s counterparts, are left craving "more," until now there has been no "more" to be had—no information about her life beyond what the book conveys and no access to her other writings. A typical blog post queries, "Where in the world is Juanita Harrison?" alleging, "We don't know what happened to her after the book was published. We don't even have a picture of her." Another reader confides, "I was totally living vicariously through Juanita. I didn't want to rush it.

And when the book ended, it left me wanting more."[13] Hungry readers have had to make do with the stock biography recycled across encyclopedias and other source volumes, with factual errors accruing along the way (that Harrison began traveling at sixteen, that no photos of her exist, that her life after publication is entirely obscure). This study offers both prequel and sequel to *My Great, Wide, Beautiful World* along with exegesis of the text.

Harrison pursued not upward class ascent but rather leisure, pleasure, and novelty. "Its my life to see and enjoy," she insisted, and her actions backed her words.[14] Adding to our store of knowledge about her helps redress the critical neglect of ordinary African American women, especially those who fall outside certain social categories. In her study of a nineteenth-century murder trial, *Hannah Mary Tabbs and the Disembodied Torso*, Kali Nicole Gross observes that most working-class Black women lacked the resources to write life narratives and that they usually appear in the historical record only when their lives "intersected, or collided in some way, with white people."[15] She notes that scholars, moreover, privilege a narrow range of subject positions: "We have come to know black women as laborers. . . . We encounter these women as problems. . . . We understand them as displaced. . . . We glorify them as freedom fighters." She concludes, "We do not have many stories of individual women who lived for themselves and did not put the race or their children or families first."[16] *My Great, Wide, Beautiful World* contributes one such story, told in the subject's own words, and this biography amplifies it. To adapt Gross's characterization of her subject as "ordinary but infamous," Harrison was ordinary but celebrated.[17] Her renown stemmed from her incongruity: that a seemingly unremarkable woman wrought such remarkable travels, and that in their course she displayed such unexpected writing talent.

Harrison's migratory history as a domestic worker sheds light on the lives of other working-class African American women of her day. Broadening the class and gender purview, her diasporic practice and transnational subjectivity also link her to contemporaries whose experiences were more institutionally structured. Harrison was not directly connected to, but did move along the periphery of, diverse intellectual and social Black movements. She published at the tail end of the Harlem Renaissance, and some of its major figures, including Langston Hughes, Alain Locke, and Carl Van Vechten, knew her book. *Ebony* magazine editor Era Bell Thompson noted that *My Great, Wide, Beautiful World* inspired her to take her first real trip, to San Francisco by train.[18] Harrison joined other African Americans, including writers, artists, performers, entrepreneurs, and activists, who through living and working in Europe could distance themselves from US racism. Her time in Paris bordered or overlapped with that of Fauset, Hughes, Josephine Baker, James Baldwin, Gwendolyn Bennett, Countee Cullen, W. E. B. Du Bois, Claude McKay, and many others. Positioning herself as a world citizen, she put into lived practice a working-class mode of the international race solidarity espoused in official venues such as the Pan-African congresses and in publications like Du Bois's *Dark Princess* and Eslanda Goode Robeson's *African Journey*. William Pickens, activist and future NAACP director of branches, made a point to visit her in Honolulu, drawn to this wanderer who enacted his ideal, as Marlon Ross expresses it, "of race progress that hinged on the physical movement of peoples across continents."[19]

As this web of connections suggests, recovering the events of Harrison's life illuminates not only her own choices but also, more generally, "Black place-making practices in the face of oppression," to quote from the seminal *Southeastern Geographer* special issue on this subject.[20] Katherine McKittrick's *Demonic Grounds: Black Women and the Cartographies of Struggle* has helped me recognize Harrison's text and practice as an example of how "black women's geographies" challenge "seemingly predetermined stabilities, such as boundaries, color-lines, 'proper' places, fixed and settled infrastructures and streets, oceanic containers."[21] Her strategic travels can be read as a scaled out, postemancipation version of the defiant acts under slavery that Stephanie Camp tallies in *Closer to Freedom: Enslaved Women and Everyday Resistance in the Plantation South*, which shows how enslaved women found a measure of liberty through laying claim to a multitude of small spaces.

Throughout her life, Harrison mobilized, to use Judith Madera's formulation, "operations of place as creative strategies for living."[22] As her travel history lengthened and her world knowledge accrued, increasingly she understood each place in relation to all those she had known before. Her personal geography of resistance comprises not so much a conglomeration of the places she visited and passed through as the relationships she established between them. By virtue of not only her onward journeys but also her many returns, imaginative as well as physical, she positioned herself within a dense spatial network.

For the most part Harrison did not leave the beaten track. Instead, she paced well-trodden tourist terrain: the Moulin Rouge, the Taj Mahal, the baths at Baden-Baden, the bullfights in Seville. Her itinerary can seem near clichéd. Yet by virtue of her practice and very presence at these sites—and by writing about it—she registered them in new ways.

Madera articulates this process in stating that "places are generated through processes of participation," that "places are about how we use them, how we share them." Representing them in print, therefore, "arranges geographic meaning."[23] She contends that African American discourse is characterized by narrative acts whereby familiar "geographic contexts and descriptors" are dislodged. "To gain representation in the symbolic structures of white territorialization," she argues, "black authors had to write over white principles of containment. They had to dismantle dominant organizational codes of place."[24]

We see Harrison making just such moves in recasting the Grand Tour narrative. Within a set of near-obligatory tourist destinations, her travelogue maps a maze of microsites—at markets, cafes, YWCAs, parks, and private homes; on board ships and in railway cars; on streets, beaches, and riverbanks—whose meanings are produced by her receptive sensibility, her interactions with other people, and the nonnormative perspective she brought to bear on them. This perspective was fostered by her racial identity and racialized experience in her home country. To use another of Madera's concepts, like so many Black writers, Harrison charts a countercartography.[25]

In quite material ways, her experiences call our attention to what diverse locations around the world could offer disenfranchised Americans such as she. For a questing African American woman at the turn into the twentieth century, her Mississippi birthplace was the absolutely wrong place. But once she left her home state behind, Harrison was usually in other places during optimal windows of time—due to her deliberate method. Prior to leaving North America, she found an accommodating home in the Republic of Cuba just as it was transitioning into a major tourist destination and entering a heady economic boom. She was in Paris at its gayest moment and in Los Angeles at its most expansive one. Across Europe, she secured remunerative positions with ease, often with other American expatriates, from whose financial reversals she profited after the stock market crashed. On her return to US territory, as an American citizen she benefited from her country's occupation of Hawaii and its developing tourist economy. Shifting to another continent, she left Honolulu prior to the attack on Pearl Harbor to travel widely in South America, and she lived for nine years in Argentina just as it became a workers' haven under Juan Perón.

Harrison broke with Mississippi but never established a single geographical center elsewhere. Instead, she valued clusters of places and the associations she made between them, along with her perpetually refreshed social contacts. She especially relished sites where different communities rubbed up against each other to produce new cultural forms—as when she is on a ship owned by one nation, worked by a crew from another, and patronized by passengers from still others. This alternative network was utterly modern and transnational, knit together by rail lines, shipping routes, and postal systems and energized by a rotating cast of human actors. She reveled in what each place had to offer while resourcefully remedying what it lacked, importing skills and insights garnered in one location to the next. And when economies or societies grew less favorable, she moved on.

Her biography, consequently, radically reorients our view of places and eras—Havana in the teens, the French Riviera in the 1920s, Honolulu in the 1930s, Buenos Aires in the 1940s—shifting the perspective from that of the middle-class or upper-class white American tourist intent on leisure to that of a working-class Black American intent on making a living—and on leisure, too. The picturesque whitewashed, brass-trimmed stoops of London take on new meaning when they are assessed as requiring too much labor to keep up. "I promised myself," Harrison remarked on seeing them, "when I Looked For a job not to get in a house with such work" (4).

Greater knowledge of her life also supports more informed readings of *My Great, Wide, Beautiful World* and its unusual rhetorical history and collaborative production. As the epistolary memoir of a working-class Black woman that reached popular and critical audiences nationwide, the book readily commands a place on the American literary historical map. The text exhibits its protagonist's pursuit of literacy and passion for mobility—two keynotes in African American literature—and demonstrates how closely they were intertwined. It thereby joins and shapes this long tradition, even as Harrison rarely comments on racial issues or calls attention to her racial identity.

Her book can be further situated within several subgenres of African American literature. *My Great, Wide, Beautiful World*, attending in equal measure to work and play, shows that the genres of labor narrative and travel narrative need not be mutually exclusive; economic migrants are as much travelers as middle-class holidaymakers. It is an exemplar, moreover, of working-class epistolary writing, comprised largely of letters Harrison wrote to former employers; the daughter of one of them served as her editor and agent. The book offers eloquent testimony about her relationships with the women for whom she worked, as much through its structure and composition as through its assertions and scenes. Its reception matters, too, for the light it sheds on the ways a mainstream national audience and more locally situated readerships registered this account of an African American woman's experience.

Stretches of Harrison's personal and authorial history still elude us. Indisputably her many physical relocations, set into motion by racial conditions but largely voluntary, did not foster recordkeeping. Yet looking past her individual choices, a more systemic cause is the paucity of records for African American cultural production. To select the most egregious example, Harrison is entirely absent from the voluminous papers of her *Atlantic Monthly* editor, Ellery Sedgwick. Her race surely contributed to this absence, given the attitudes prevalent across white print culture and, more personally, Sedgwick's racist inclination to conceive of Harrison as first and foremost a servant, not an author.

Any archive, Carolyn Steedman reminds us in *Dust: The Archive and Cultural History*, is a manifestation of institutional and state power, "selected and collected rememberings and forgettings." Eric Gardner attributes the thinness of African American archives to "the grasp of white power on much of American print culture and American memory." Complementing Gardner's observation that Black literary production was often viewed as "ephemeral," John Kevin Young argues that the dearth stems from a conception of it as document rather than art, rendering superfluous any preservation of its material iterations. "We must first deal with histories riddled by silences," Madera states in respect to uncovering Black geographies. Scholars working in all periods of African American geography, history, and literature, among many others Lois Brown, Frances Smith Foster, Marisa J. Fuentes, Saidiya Hartman, Laura E. Helton, Joycelyn K. Moody, and Pat Noxolo, model an array of tactics to "deal with" such silences. Collectively they demonstrate, in Brown's words, "the strategic and necessary ways in which we must read what we have" in the face of "rhetorical ruptures and biographical caesuras."[26]

I first began researching Harrison in the late 1990s and have already written a great deal about her, in the context of region, domesticity, and the *Atlantic Monthly*.[27] Yet I always had more to discover and more to say, a condition that has only intensified with the growing number of digitized public records and the donation of a cache of Harrison's private letters to the University of California, Los Angeles. This study allows me to both consolidate new material and recast past findings to create a new kind of biography, one that puts a bestselling travel book and previously unknown archival documents in conversation with each other.

In accord with the tenets of microhistory, I use my literary training to parse the sparse record of Harrison's life, pressing fragmentary evidence to extract the last drop of significance. Jill Lepore describes such a task as tracking "elusive subjects through slender records" in pursuit of "small mysteries."[28] Early influences include studies such as Laurel Thatcher Ulrich's *A Midwife's Tale: The Life of Martha Ballard, Based on Her Diary, 1785–1812*; like Ulrich, I render a subject's memoir the backbone of a biography even as I analyze the memoir itself. Ava Chamberlain's biography of Elizabeth Tuttle has been another, aptly described as a model for "thinking through the challenges of working with incomplete, inconclusive, or absent archives."[29] Tuttle, the notorious grandmother of Jonathan Edwards, left no written traces, but Chamberlain distills her life by examining her environment and extended family along with her position in Puritan historiography.

As I moved deeper into the study, increasingly I followed the lead of the many scholars in African American studies who recover Black women's experiences. To take one spectacular example, in *Wayward Lives, Beautiful Experiments: Intimate Histories of Social Upheaval* Saidiya Hartman transfigures irregular source materials to express the sensibilities of rebellious young women. She dubs this method "close narration" and, in an earlier essay, "creative fabulation," whereby she "elaborates, transposes, augments, and breaks open archival documents so that they might yield a richer picture."[30] Lois Brown's magisterial biography *Pauline Elizabeth Hopkins: Black Daughter of the Revolution* has been another humbling inspiration. Brown both charts Hopkins's life and oeuvre and ranges beyond them to shed light on numerous literary, historical, and political topics. Turning to a dedicated literary study, Ayesha K. Hardison has alerted me to the diverse ways that mid-twentieth-century writers construct Black female subjectivity, a process that she identifies as "writing through Jane Crow" in her book of the same name. Hardison explains that "the moniker Jim Crow did not address the patriarchal conventions also subjugating African American women." Instead these women grappled with what she terms "Jane Crow," "the burdening of black female subjectivity under a specific set of social conditions: mass migration, changing gender relations, class anxiety, and racial strife."[31] Writing a generation earlier than Hardison's subjects, Harrison, too, endured this kind of subjective strain.

Race, the legal scholar Ian Haney López states, is "neither an essence nor an illusion, but rather an ongoing, contradictory, self-reinforcing process subject to the macro forces of social and political struggle and the micro effects of daily decisions."[32] In discussing Harrison's racial position, I try to preserve a balance. I regularly highlight her identity as an African American woman and situate her in African American contexts. This is a biography of a Black writer. Yet at the same time, I respect her decision—at least in her book and possibly in her lived experience—to prioritize additional subject positions and keep some distance from African American communities. This study honors her delight in dislodging fixed identity categories while also assessing how race shaped her travels and authorship and acknowledging it was her coming of age as a Black woman in Jane Crow Mississippi that makes her achievement so immense.

The core of this project is a deep sounding of *My Great, Wide, Beautiful World*, as both biographical source material and literary artifact. An exceptionally dense record, the travelogue rehearses Harrison's movement and labor between 1927 and 1935 and expresses the philosophy that informed her decisions. It is remarkable just how much hard biographical fact a close study of the book yields, about events not only within but also outside its chronological scope. Should we apply processes of back formation and extrapolation, moreover, the text can indicate what Harrison had done in the past—in similar situations elsewhere—and by virtue of educating us on her preferences and tactics suggest her future choices as well.

This exploration is further anchored by critical readings of a group of key documents. They include Harrison's revealing letters to her friend Alice Foster in Los Angeles, admiring Honolulu newspaper profiles at the height of her fame, a missive she fired off about the value of *Uncle Tom's Cabin*, and a fortuitously detailed application to renew her passport at the United States Consulate in Buenos Aires. Small but expressive, this collection underpins the study; indeed, without the Foster letters I scarcely could have written it. I also of course draw upon public records, including censuses, city directories, passenger manifests, immigration cards, hotel registries, probate documents, and visa, passport, and Social Security applications. I look too at what we learn from the actual architecture of Harrison's life, the buildings in which she lived and worked. Throughout, rather than naturalize the life-writing process I specify the evidence I use. I never want readers to ask themselves, how does Halverson know this?—to wonder whether invention, extrapolation, or actual documentation is the source of a claim.

The public discourse in which *My Great, Wide, Beautiful World* is embedded also has much to tell us, as comprised by editorial commentary, book reviews, author profiles, readers' letters, book club minutes, and memoirs. Lloyd E. Ambrosius's counsel well applies to a study of Harrison: "When, as is often the case with women or people of color, primary sources are inadequate . . . biography must make creative use of whatever is available."[33] Yet using such sources poses a challenge of its own, since so much discussion of Harrison reproduced stereotypes of African American women, famously characterized by Hortense Spillers as "a locus of confounded identities, a meeting ground of investments and privations in the national treasury of rhetorical wealth."[34] Thus when perusing mainstream representations of Harrison, the information they offer must be sifted from the bias they express. Leon Jackson formulates such a task as "reading through the lines and against the grain" of what he terms "compromised sources."[35]

The compromised nature of much of the commentary about Harrison, as tendered by white employers, editors, journalists, and reviewers, makes it all the more critical to quote extensively from her public and private writings, so that she be represented in her own words. Her rationale for settling in Honolulu, as a "lovely part of the World" (314) that recollects the peace, beauty, and cultural richness she savored on the French Riviera, is very unlike the explanation given by her patronizing editor, that in Hawaii "a minimum of work gains abundant food" (xii). Harrison herself, however, is not always a reliable guide. Assertions that she makes in her book and her private letters can contradict each other, and when it served her she dissembled. She regularly took years off her age, sometimes claimed to be married, and, of greatest interest, had a penchant for "passing," not as white but as non-American and, often, as a person of higher or lower class. In both their lives and their discourse human beings engage in many forms of identity performance, and not all of them are racial.[36]

Aside from that two-year burst of celebrity at almost exactly its midpoint, Harrison had an inconspicuous life. It was structured, however, by time residing in world capitals, visiting famous sites, attending illustrious events, and consorting with cosmopolitan populations. In declaring "I want alway to be where wealth health youth beauty and gayness are altho I need very little for myself I just want to be in the midst of it" (318)—a statement that we might take as her life credo—Harrison marks the incongruity between her humble lifestyle and often glamorous environs. She lived in what had once been Mumbai's legendary Watson's Hotel. She composed her book in a borrowed apartment with a view of Notre Dame. She worked among the bohemian artists who gathered in southern France. Living in her custom-designed tent, she was just a stroll away from Waikiki's iconic, bright pink Royal Hawaiian Hotel. In Madrid, we can track down the exact bullfight she saw and its star toreador and identify the Diego Velázquez painting she contemplated in the Prado Museum while on a nursemaid's outing with her two-year-old charge. Her haunting of storied places makes her experiences easier to recover and their contexts more readily comprehended. So too does the fact that many of the people she associated with were prominent figures with recoverable profiles of their own. Her editor Mildred Morris, for example, had once been a famous child actor, starring as Wendy in the first Broadway performance of *Peter Pan*.

Lingering over the places Harrison dwelled, this book also introduces readers to the American families—Black, white, and Asian—who influenced her life course, namely, the Criglers, Fosters, Dickinsons, Morrises, and Tadas. As some compensation for what we do not know about Harrison, I proffer as much pertinent detail about them as I can muster. There is a paradox here. As a child Harrison was separated from her family with no apparent regret. Once she left Mississippi she never returned. She was smug about being single. She made it a rule to travel alone, and her stated intention was to move on before her friends could bore her. She insisted, "I planned my life and in it I did not take neither man Woman nor Child" (FL2). Yet contrary to Harrison's own inclination, this study privileges her human connections. Proffering fine-grained accounts of these now unknown families and their own migratory histories is one of my tactics to uncover her experience and present the texture of her life.

Ranging across four continents and 130 years, from the 1830s to the 1960s, the eclectic collection of topics in this book—the likes of Los Angeles evangelist Aimee Semple McPherson, the Muroto typhoon, the history of the YWCA, ethnic hierarchies in Hawaii, "Negrophilia" in France, and many others—reflects the eclecticism of its subject's itineraries and interests. Harrison's peripatetic trajectory wreathes countless locations and the events that took place there, along with countless histories, institutions, personalities, and movements. Any biography of her, consequently, becomes an alternate way to organize eighty years of world history. Her life exhibits the interplay of global historical forces and their impact on a single individual in ways that are rarely so humanly apparent. Finally, less loftily, I hope that touring the constellation of physical and social landscapes that Harrison assembled across much of the twentieth century will delight the readers of this book just as they once delighted her. Following the stages of her life provides a way for us to take a world tour, too, visiting enchanting places in her ardent company.

The book is divided into three sections that chart Harrison's life before, during, and after the span of time that *My Great, Wide, Beautiful World* covers. Throughout I look to address her motives, methods, and modes of expression and self-presentation; probe her psychological influences and provide social contexts; identify the personal strategies and larger systems that enabled her to work, travel, and write; and show how she transformed the raw material of her experience into her ebullient book.

Part I, Before *My Great, Wide, Beautiful World*: American Beginnings, opens with a chapter on Harrison's upbringing in the South. In addition to tracing her multiethnic family origins—African, white, and possibly Choctaw—the chapter shows how two opposing trajectories structure her first twenty-one years: her individual arc of increasing agency and, in tension with it, the process by which her home state shut down the rights of its Black residents. Chapter 2 follows Harrison through the roving across North America that constitutes the first major stage of her itinerancy, tracking the places she inhabited, the labor conditions she navigated, and the strategies she devised to thrive in hostile geographies.[37] Over the course of seventeen years she resided in every US region and at least sixteen states, as well as Cuba, British Columbia, and the territory of Hawaii. This wide-ranging North American tour is followed by a chapter that zooms in on her time in Los Angeles in the mid-1920s and her relationships with the Black and white Californians who energized her world journey and voluble correspondence alike.

Part II, During *My Great, Wide, Beautiful World*: World Travel and Authorship, turns to *My Great, Wide, Beautiful World* and the period in Harrison's life that it maps. Along with using Harrison's travelogue as biographical source material, I offer a critical sounding of the text itself. I assume that most readers of this study have read, or are soon to read, Harrison's book. Nevertheless, I quote from it liberally, both to have Harrison narrate her experiences in her own words and to analyze its arguments and themes. In these quotations I avoid resorting to ellipsis or other amendment, even at the expense of smooth integration, so as to more accurately represent her grammatical structures and trains of thought.

Chapter 4 begins by untangling Harrison's web of travel itineraries and employment histories, whose sheer multiplicity can be confounding. Entering into a more literary reading, it then examines the connections *My Great, Wide, Beautiful World* makes between writing, travel, and freedom and its demonstration of how Harrison's multiple geographical locations allowed for multiple ethnic, class, and national identifications; it also explores the gender arguments that thread through the text. Chapter 5 focuses on Harrison's nascent authorship in conjunction with her four-year residency on the French Riviera. It traces the collaborative process whereby she composed the book in Paris and investigates her proximity to the so-called Lost Generation, between-the-wars writers who—just like her—found inspiration in the creative ferment and gorgeous landscapes of Provence. The last chapter in this section returns to the travel record to follow the final stage of her world tour and assess her experiences as a racially marked Western traveler, homing in on events in North Africa and India.

Part III, After *My Great, Wide, Beautiful World*: Celebrity and Beyond, is a reckoning of Harrison's life after publication. It opens with twinned chapters that survey her mid-1920s and mid-1930s sojourns in Hawaii, followed by an overview of the reception of *My Great, Wide, Beautiful World*, which became a point of reference for Honolulu boosters, source of pride for Black readers, and fount of jollity for white women's book clubs. To better account for Harrison's relatively undocumented South America period in the 1940s, chapter 9 departs from a strict chronological progression to investigate her experience around the pan-Hispanic world. Ranging back in time to gather up her youthful enterprise in Cuba and midlife activities in Spain, it then turns to her travels across South America and decade of residence in Argentina. The final chapter sketches her last years, first in La Jolla and Venice, California, and then in Honolulu once again. At the close of her long life, Harrison was identified as white.

The epilogue returns to her print culture activity in probing the implications of the mid-1940s letter Harrison sent to drama critic John Mason Brown. With this overture, the latest extant text we have in her hand, she entered the national debate about the value of Harriet Beecher Stowe's abolitionist novel *Uncle Tom's Cabin*. Opining on a book about the South prompted dark childhood memories and an urgent appeal for greater African American representation.

Harrison had supreme confidence in the depth, worth, and appeal of her life history. In a private letter she depicts Morris as rightfully fascinated by the memories she shared as they worked on the book manuscript together in Paris. She states, "At the begining the Writer thought that she would type a short sketch of my life for the interducting of the letters. but I knew that it couldn't be a short sketch. because my life are like the Saying to 'never judge a book by its covering' sometime the covering look not at all enviting but the enside are most interesting." She adds, "She became so interested that she said she could not do Her other writings she was so anxious to get down to me to go on with my true life story" (FL2). Taking up her metaphor, we can ask why Harrison thought that the "covering" of her life belied its magnetism. Because of her occupation? Her gender? Her language? Her race? Despite their "interest," she implies, the histories of people like her were rarely told. Now, however, due to her and Morris's belief in its value, one such story would be broadcast. Indeed, inspired by their energized exchange, they laid plans to publish a full biography. That project never came to pass, and the granular oral record Morris transcribed has been lost—condensed to a bare outline in the preface to Harrison's book, a "short sketch" after all. While that loss is irrecoverable, with this book I tender an alternative version of Juanita V. Harrison's "true life story."[38]

PART I

Before *My Great, Wide, Beautiful World*

American Beginnings

CHAPTER I

Mississippi Ambition

In a July 1927 entry about her first day in Dijon, France, Harrison describes her experience at a movie theater:

> I went to a movie to see Mary Pickford in Sparrow. The girl said the Programs was one Franc. So I did not buy one and she said if that was too much for me I could read it then give it back to Her so I did. just in frount of me sat a French Lady that spoke English they bought candys and gave me some and the time passed very plesant. (39)

The inventory of tactics to optimize a leisure experience is typical of *My Great, Wide, Beautiful World*: paring expenses, navigating language difference, socializing with local women, scoring free food. Likewise, that Harrison offers no evaluation of the film itself—which features destitute children fighting for their lives in the American South—accords with her usual silence about her native region. Her book and correspondence give only clues about her early life in Mississippi. This chapter seeks to assemble those clues, along with other source materials, into a fuller portrait.

Pickford's biographer characterizes *Sparrows* as "redolent of mud, muck, snaky vines, staleness, dust and heat" but having "a strange, almost lurid beauty" nonetheless.[1] It portrays the travails of plucky Molly, who is held captive with eleven younger children at a "baby farm" in backwoods Louisiana, where they are worked and starved nearly to death by the sadistic Grimes family. Molly and the others escape into the quicksand-riddled swamp and manage against the odds to survive. Most of the film's action takes place in an expanse of mud, such a stark set that child welfare officials were called in to investigate.

We can read the entrapment of these white foster children in Louisiana as a grotesque metaphor for Harrison's constriction as a Black orphan in Mississippi, a motherless poor girl growing up in the South. She was mired in an unjust, oppressive society, at the mercy of white individuals and institutions and living with the specter of racial violence, which became even more ominous as she was coming of age. At the same time, however, *Sparrows* speaks to the fierce determination that she mustered in response. Like Molly and the children she shepherded, Harrison, too, found a way out of her personal swamp. Even as a girl, she began to realize her goal of living in a way utterly unlike those around her, rich or poor, Black or white. "I planned my life at 11," she reveals in a letter she wrote to her friend Alice Foster from the vantage point of her mid-forties. "At 13 I begain to live it a little like I had planned then at 21 I had it just right and at this moment am enjoying those planes" (FL2). The skeleton chronology also includes being "born in spirit" at seventeen.

Scarcely two months after Harrison's actual birth in the northeast Mississippi town of Columbus, the Mississippi legislature segregated the state's trains. The decision reflects the increasingly rigid racial caste system that structured her life, a system that controlled not only the physical but also the economic and social mobility of African Americans. Two seemingly contradictory trajectories run through Harrison's youth: her personal arc of enlarged freedom and personal fulfillment and, in tension with it, the sequence of acts by which her home society, reversing the gains made in Reconstruction, sought to confine its Black population.

The application to renew her passport at the US consulate in Buenos Aires identifies Juanita Harrison as an American woman of five foot three, with brown hair and brown eyes, daughter of Jones Harrison and Rosa Crigler. The attached photo shows her to have broad shoulders, a piercing gaze, a light skin tone, and wavy hair, parted in the middle and tied back. (Her smile is too slight to corroborate the statement she makes in her book, that she had abundant gold fillings in her teeth.) The application correctly gives her date of birth as December 26, 1887, as do all her other government documents. Unofficially, however, Harrison often shaved four years off her age, leading to erroneous assertions such as her editor's claim that she hit the road at the tender age of sixteen.

The more formidable barrier to reconstructing her and her family's early history is of course the paucity of records for ordinary African Americans. To quote Frederick Douglass, "Genealogical trees do not flourish among slaves."[2] Information about their histories, families, and residencies is found not in the census but in the cursory detail of slave schedules, mortality schedules, and estate records. Marisa Fuentes states that to truly recover nineteenth-century

African American experience, "One must persist for years in this 'mortuary' of records to bring otherwise invisible lives to historical representation."³ The convention of not fully naming enslaved individuals or delineating their family relationships, the cataclysm of the Civil War, and the resulting population shifts in Mississippi and the rest of the South foreclose definitive knowledge about Harrison's ancestors. Add to this her childhood upheaval: After the death of her mother, she was separated from her family "before I were old enough to know all their names," as she put it.⁴

My Great, Wide, Beautiful World sheds little light. While the book is peppered with allusions to Harrison's previous American dwelling places, Mississippi is not among them, and with one important exception it does not discuss events from her youth. Over the course of her travels Harrison spun an ever denser web of references, with each successive destination helping her evaluate the others. Yet the South does not have an evident place in this network. Patricia Yaeger observes, "The South is most conspicuous in its absence from her story: a travelogue in which almost every other place on Earth gets visited, celebrated, named."⁵ As we will see in a later chapter, the region does in fact surface in her text, but in ways that she does not seem to control.

Harrison had a large extended family whose members lived in close proximity to each other. She never mentions her father or his branch of the family. All we know about the man is his name, Jones Harrison, his birth year, 1862, and his birthplace, Columbus. Possibly her parents were not married, as she did not identify her mother as a Harrison on official records. Given that she makes no mention of brothers or sisters, likely Juanita either was an only child or had little contact with her siblings.

Happily we know much more about her maternal relatives, the Criglers, a name that in accord with her flexible orthography, Harrison variously spelled Creglar, Crutler, and Creagler.⁶ The 1870 federal census shows her future mother, Rosa (spelled Rose), at age three residing in Choctaw Agency, Oktibbeha County, with her own mother Lizzie Crigler and a sister and two brothers (Ester, Joseph, and Walter, aged six, five, and one, respectively).⁷ Choctaw Agency, straddling an east-central Mississippi crossroads, was a trading post founded in 1813 to help the US government establish ties with Native communities.⁸ By 1860, with the majority of the Choctaw forced westward, its function had narrowed to that of a busy stagecoach stop. While the western part of Oktibbeha County was hilly wooded land tilled by its owners, the wealthier, more fertile farms to the east were worked by enslaved women and men.⁹ The county's population that year was enumerated at 5,328 white people, 7,631 slaves, and just eighteen "free coloreds."[10] Proof of the community's terrifying racial politics, at the time of Lizzie Crigler's residence, Choctaw Agency was home to an active branch of the Ku Klux Klan. After emancipation, many of the county's African American residents who did not leave the state altogether had moved to adjacent Lowndes County, Harrison's birthplace.[11] Columbus's population increased by 50 percent in the five years after the war.[12]

The 1870 census identifies Lizzie Crigler as Black and all four of her children as "mulatto." It discloses that she worked as a cook, could not read or write, and was born in 1840 in North Carolina. What brought her to Mississippi? Most likely, she arrived enslaved. Traveling either overland or by sea, she may have been taken from North Carolina by an owner, or she may have been sold from the Upper South into the Deep South. Precipitated by the wholesale removal of Native populations, the Southeast saw a proliferation of new plantations in the mid-nineteenth century, and the domestic slave trade swelled to meet new labor demands. In *Black Slaves, Indian Masters: Slavery, Emancipation, and Citizenship in the Native American South*, Barbara Krauthamer explains, "The land that had once been Indian country quickly became the United States' 'kingdom' of cotton, as white slaveholders and black slaves flooded the region. From the 1830s until the Civil War, hundreds of thousands of enslaved African Americans were sold from the Upper South to work and die on the mostly white-owned plantation labor camps that covered former Indian towns and hunting grounds."[13] The magnified scale of production ushered in an era of great prosperity for the planters, landowners, and merchants who comprised the state's white elite. Mississippi became the nation's leading producer of cotton and third largest slave-holding state, with over four hundred thousand enslaved people making up 55 percent of its population.

Of German ancestry, up until the 1830s America's Criglers were mainly clustered in Virginia, Kentucky, and Missouri. However, by 1880 there was a preponderance of them in east-central Mississippi, too, where their descendants live to this day. The Mississippi Criglers can be traced back to two Virginian brothers, Jason C. Crigler and John A. Lewis Crigler, who moved to the district as young men on the heels of the 1830 Treaty of Dancing Rabbit Creek. Following on Andrew Jackson's Indian Removal Act, the treaty compelled the Choctaw to cede more than ten million acres of tribal lands in eastern Mississippi and western Alabama. In exchange they were granted land in Indian Territory in present-day Oklahoma, and 80 percent of the population was forcibly resettled.[14] Sons of Lewis Crigler, a prosperous miller and slaveholder in Virginia's Culpepper County, Jason and John were among the thousands of white settlers who hurried westward to buy vacated Native lands from the federal government, settling in different towns in 1833. John, at least, is reputed to have brought enslaved people with him. His son, Dr. John Lewis Crigler, became one of Oktibbeha County's most prominent residents. During the Civil War, he was made first lieutenant of the Fourteenth Regiment of the Mississippi Infantry, and he served in the state senate from 1880 to 1884.[15]

During Harrison's youth, many Criglers, Black and white, lived in the area. An Oktibbeha County genealogist indicates the origins of her family name in reporting, "I have found that many blacks in the county DID take the last name of their owner."[16] In a private letter, Harrison refers to "my own Kin which I have many" (FL2), and Korey Garibaldi maintains that she likely knew the identity and "whereabouts" of her white maternal grandfather and other white relatives.[17] A present-day descendant states, "Oktibbeha County Mississippi was once saturated with black Criglers until they loaded up and moved to the delta in search of a better life." Another self-identified "Black Crigler" notes that she is "descended from a branch who moved from Mississippi to Oklahoma around 1890," a migration that occurred during Harrison's early childhood.[18] The next generation saw the evocatively named Pocahontas Crigler.

Beyond revealing that he was not Black, the 1870 census offers no information about the father of Lizzie Crigler's children, Harrison's grandfather. If the same man was father to them all, then Lizzie's relationship with him commenced prior to emancipation and endured for some years. Her eldest child, Ester, was born in 1864, and the other three by 1869. It may of course have continued long past then.

One paternity candidate is farmer Benjamin Parish, Lizzie Crigler's neighbor. Parish, one of only five white men in the 1870 census of Oktibbeha County Beat 5, was employed by Green B. and Lucy Stallings, wealthy landowners. Indicating the proximity of their homes, his listing appears directly above Crigler's in the census registry of households. Single in 1870 and just two years older than her, Parish was also born in North Carolina, and he had briefly fought for a North Carolina regiment prior to coming to Mississippi. Perhaps Crigler, whose occupation is listed as that of cook, also worked for the Stallings. Among other connections between the Stallings and the white Criglers, the aforementioned Dr. John Lewis Crigler was married to Green and Lucy's daughter Lydia. Lizzie Crigler may have even grown up in the Stallings's household. A county slave schedule inventories the twenty-seven human beings to whom Green B. Stallings had legal title in 1860. Ten of them were children, including a girl of ten, unnamed, the same age Lizzie would have been at the time.[19]

Regardless of whether the children's father was Parish or some other local man, the nature of his and Crigler's relationship is unrecoverable. It might have been a nurturing union between a white man and a Black woman who were unable to legally wed. It might have been an exploitative arrangement that Crigler entered into against her will. It might have had elements of both, a blend of coercion, affection, and mutual dependence.

The Criglers' geographical location also allows for an alternative origin story, one in which color lines are even more blurred and racial mixing began generations before Lizzie's children were born. The same racially structured events that brought the white Criglers from Virginia to Mississippi—the forced removal of Native Americans and the expansion of plantations worked by slave labor—brought together African and Native people. Harrison maintained that she was part Native American, and her grandfather might have been among the four to six thousand Choctaw who continued to live in Mississippi after the tribe ceded its lands.

The territory's Native and African residents shared a dark history. The Choctaws had been prosperous traders and farmers in Mississippi and Alabama, and the more affluent owned slaves. Since traditionally Choctaws held land communally, the institution of slavery gave individuals a way to invest in private property.[20] The Choctaw Nation backed the Confederacy. Krauthamer eloquently states, "The histories of chattel slavery and Indian removal . . . are not simply parallel or even competing narratives but are intertwined histories of destruction and dislocation."[21] Forced westward to Indian Territory, Choctaw and Chickasaw tribal members took with them hundreds of enslaved people; some had sold treaty-granted lands in order to purchase mobile living property.[22] Many were already culturally Choctaw—speaking the tribal language, adopting Choctaw rituals and everyday customs. After emancipation, they and their descendants were known as freedmen. Some succeeded in entering tribal rolls before they closed in 1907, the fruit of much advocacy and petition. The Black Criglers who migrated to Oklahoma in the 1890s were probably motivated by Choctaw ties.

Were the father of Crigler's children Choctaw, he was likely of mixed race himself. Commencing in 1780, there was extensive mingling between Mississippi's Native and white (and to a lesser degree, African) populations, especially the French. Mixed-race families were disproportionately represented among the Choctaw who did not leave Mississippi, as more had the financial resources and social integration with white communities that enabled them to stay.[23] They made up a high proportion of successful claimants under Article 14, the codicil of the

Treaty of Dancing Rabbit Creek that allowed Choctaw men and their families to file for allotments of tribal land and receive US citizenship.[24] In contrast, many of the full-blood Choctaws who remained in the state withdrew to its east-central sandhills, where they lived in relative isolation at subsistence levels.

Although she is easily overlooked, the sixth person enumerated in Lizzie Crigler's household, boarder Manerva Nail, further encourages us to weigh the possibility of Harrison's Native ancestry. The census tells us that Nail was eighteen, Black, and a servant. But her actual name reveals something more, as Nail was a common surname among the Choctaw.

The first white settlers in Choctaw country included a number of English Nails, who through establishing Native family ties attained considerable influence and prestige.[25] Revolutionary war hero Henry Nail, for example, had a Choctaw wife, and their daughter Rhoda Nail married into the powerful Folsom family, whose members were early tribal leaders.[26] Joel H. Nail was one of the mixed-blood signers of the Treaty of Dancing Rabbit Creek, and he, Molly Nail, and Robert Nail all received land through Article 14.[27] Some Nails were slave owners in Mississippi, and a branch of the family took one Joe Nail, enslaved, with them west to Indian Territory.[28] The Choctaw Final Rolls list thirty-three Choctaw Freedmen Nails along with fifty-three Nails and Nales who were Choctaw "by blood."[29] The name Minerva Nail (with an "i") also recurs in the Rolls. Joe Nail and his daughter Minerva are among the freedmen enumerated there, although the latter is not the Nail who lived with Crigler.

Manerva Nail, then, eighteen years old in 1870, could have been a freed woman once owned by Choctaws as a child, or she could have had mixed ancestry. Her presence in Lizzie Crigler's household points to Crigler's links to Choctaw society and increases the odds of her having had a relationship with a Choctaw man. We should bear in mind, however, that in the mid-nineteenth century the Constitution of the Choctaw Nation had strict anti-Black measures, and mixed marriages were forbidden. A Black woman's relationship with a Native man was not necessarily more equitable than a relationship with a white one.

While Harrison does not voice any claim to Native ancestry in her book, she alluded to it after publication. Honolulu journalists characterized her as an "American Negro-Indian author," as "a woman of negro and Indian blood," as "Part American Negro, Part American Indian and Part just plain American."[30] Given that the profiles drew upon personal interviews in Honolulu, Harrison herself was the source of the information. She may have been passing on verified family history. She may have relayed family lore that she believed to be true. Upon consideration of the collected evidence—Lizzie Crigler's Choctaw Agency household, mixed-race family, and boarder Nail; the emigration of a branch of "Black Criglers" to Oklahoma; Harrison's statements—there is certainly a case to be made. At the same time, given her bent for invention, we cannot rule out the possibility that she made it up, perhaps as a bid for status in circles where a measure of Indian blood was seen as alluringly exotic, or perhaps even as a bid to align herself with Indigenous Hawaiians. Speaking to the prevalence of such claims, in her 1928 autobiographical essay, "How It Feels to Be Colored Me," Zora Neale Hurston wryly comments, "I am the only Negro in the United States whose grandfather on the mother's side was *not* an Indian chief."[31]

And finally, what to make of her Spanish first name, Juanita? The name has prompted speculation about Mexican ancestry, especially given the relative proximity of Mississippi to Mexico and the state of Texas, which had once been Mexican territory. Yet in fact it situates Harrison squarely within mainstream US culture. In 1887, the year of her birth, Juanita was a surprisingly common name for American girls, and it became even more so in the first decades of the twentieth century, competing with other faintly exotic non-Anglo names such as Isabella and Charlotte. Moreover, it was a popular choice for Black parents. Issues of the *California Eagle*, one of Los Angeles's African American newspapers, record the doings of countless youthful and matronly Juanitas. In her essay about African American women's journals, Chimene Jackson refers to Harrison as having a "'Black' name."[32] Yet while not a sign of actual Hispanic heritage, being called Juanita facilitated Harrison's expatriate experience in Cuba, Spain, and South America. "My name are Familer to them" (174), she remarked, with appreciation, in Spain. And for the biographer, it is a boon. In 1887, the year of her birth, the top three names for newborn girls were Mary, Ann, and Elizabeth. An "Ann Harrison" would have been far harder to track.

Making her personal achievement all the more extraordinary, Harrison came of age in Mississippi just as the rights of the state's Black residents were systematically foreclosed. 1877 marked the official end of Reconstruction with the withdrawal of federal troops from the South. A decade later, the segregation of Mississippi's trains ushered in an era in which Black and white people were kept separate in nearly all aspects of life, a practice sanctioned nationwide by the ruling in the 1896 Plessy v. Ferguson Supreme Court case, which stipulated "equal but separate accommodations for the white and colored races." The decision, to quote the sole dissenter, consigned African Americans to a "brand of servitude and degradation."[33] 1890 saw the passing of the Second Mississippi Plan, whose raft of literacy requirements and poll taxes resulted in a 90 percent drop in

African American voting. Anti-Black violence accelerated, and Lowndes County was especially afflicted, suffering the third most lynchings in the state.

The "fundamental commitment" among the dominant social order in the South was that of keeping Black residents "in their place," as forms of institutional control were recast after emancipation.[34] Stephanie Camp states that during slavery, "Enslaved people's inferior and subjected position within the framework of antebellum southern society, their social 'place,' was reflected and affirmed by white control over their location in space, their literal place."[35] She refers to these structures as "geographies of containment," and Jim Crow policies created new ones for a new age. Mississippi actually had fewer "black codes" than did other Southern states; the policing of African American mobility was so sweeping and the punishment for breaching the racial order so severe as to make them superfluous. In her narrative history of the Great Migration, *The Warmth of Other Suns*, Isabel Wilkerson arrestingly figures this system as "an invisible hand" that "ruled . . . the lives of all the colored people in Chickasaw County and the rest of Mississippi and the entire South." "It wasn't one thing," she states, "it was everything."[36]

Harrison's hometown of Columbus had crucial functions in the state's racially structured economy. Originally a trading post, in 1830 it became the county seat of Lowndes County, serving as an administrative center to process the enormous transfers of ceded Native lands and hold hearings for Article 14 of the Treaty of Dancing Rabbit Creek. As the cotton plantations surrounding it burgeoned, the town became an important collection and distribution point for the crop. It housed both a hospital and an arsenal during the war, but a successful defense against a Union attack preserved its city center. By the turn of the twentieth century, it supported a busy Black entertainment and commercial district, which later became a blues music hub. Columbus had distinct African American neighborhoods, most notably Sandfield, which was settled by freed people and could have been home to Harrison's family.[37]

Some of the planters' new wealth went toward the construction of opulent homes, and its eclectic mansions are now Columbus's leading tourist attraction. They bear aspirational monikers such as White Arches, Twelve Gables, and Shadowlawn. The Italianate White Arches, to take one example, was built in 1857 for cotton planter and Mississippi militia brigadier general Jeptha Vining Harris. Sporting a three-story octagonal tower, it stayed in the reputedly "proud and aristocratic" Harris family until 1967.[38] Harrison was said to have worked for wealthy women in Columbus. Did she toil in one of these splendid homes? Or, from the perspective of a poor Black girl in the South, did "rich" simply mean white and middle class? Regardless, she was not impressed. As Mildred Morris—somewhat inelegantly—ventriloquizes her thoughts in the preface to her book, "No one is there for me to copy, not even the rich ladies I work for" (ix).

What we know about Harrison's childhood comes from two key sources. One is Morris's preface, which includes an abridged, editorialized version of the oral history that Harrison relayed to her in Paris. The other is the voluble letter, previously quoted, that Harrison wrote in 1931 from Nice to Alice Foster. Foster was an African American woman from Mississippi who settled in Los Angeles, and chapter 3 discusses Harrison's relationship with her.

Harrison sat for many interviews with Morris while they worked on the book in an airy room overlooking the Seine, an enterprise traced in chapter 5. Reporting on Morris's riveted response to her "true life story," Harrison recalled, "I enjoyed watching how Her beautiful brown eyes sparkled as I went from Six years old up to Jan. 1931" (FL2). In the preface, Morris's tone can be patronizing and her assertions, racially motivated. Her description of Harrison's appearance, for example, which casts her as girlish and gestures to mixed heritage, reads as a bid to narrow the gap between the author and a white readership: "Her slight form, fresh olive complexion, long hair braided about her head, made her appear younger than her years" (xi). Nonetheless her brief overview of Harrison's childhood, below, is invaluable:

> Born in Mississippi, she had a few months of schooling before she was ten. Then began an endless round of cooking, washing and ironing in an overburdened household,—labor that might have daunted a grown person.
>
> But the child at work, clothed in a woman's cast-off apparel, stiff basque bodice, long skirt and laced bicycle boots, lived with a bright vision of templed cities in foreign lands which she had seen pictured in the stray pages of a magazine.
>
> Out of the sordid life that colored her early years she distilled a resolution: "I will sail far away to strange places. Around me no one has the life I want. No one is there for me to copy, not even the rich ladies I work for. I have to cut my life out for myself and it won't be like anyone else." (ix)

Leaving school and starting work in and of itself may have sparked Harrison's desire for travel, as the child worker recognized that she wanted something other than a life of service in small-town Mississippi. Her ambition was given visible form in the magazine she chanced across, which conjured "a bright vision of templed cities." Another version of Harrison's claim to have charted her life course at eleven, the event is the keystone of her personal myth of origins, the source of her desire to become a traveler and perhaps a travel writer, too.

The magazine may have been an issue of *Century Illustrated Magazine* or *Harper's Bazaar*, the most popular illustrated periodicals of the day. The article looks to have been a feature about Asia, given the allusion to "templed cities." (The Orientalist language employed to express her dreams is intriguing, although we can't know whether its source is Morris or Harrison herself.) Given that a magazine subscription was a middle-class luxury, her exposure was surely in an employer's home. We can imagine the young Juanita stealing some moments from her duties to leaf through "stray pages."

In figuring the beguiling scene as something "she had seen pictured," Morris implies that only the drawings or photographs made the impression, not the text they illustrated. In light of the rudimentary education meted out to Black Mississippians, Harrison may not have been able to readily read a magazine article—still at "the threshold of literacy," to use Leon Jackson's term.[39] Her employment likely contributed more than her schooling to her print culture access and emergent international outlook. Mississippi outlawed biracial schools, and of all the Southern states, it spent the least on educating its African American youth. In 1900, they accounted for 60 percent of the state's school-age population but received only 19 percent of its school funding.[40]

Education was considered especially superfluous for Black girls, slated for futures as cooks, housekeepers, maids, nurses, and laundry workers. In some ways the town of Columbus promoted women's mobility and class advancement, as it was home to Mississippi University for Women, the nation's first public women's university. In advocating for it, legislator Wiley Nash had pleaded, "Can we not do something for the poor girls of Mississippi[?]"[41] Tuition was free, each county had a quota of guaranteed places, and students could work off their board. Eudora Welty—another Mississippian who made her authorial debut in the *Atlantic Monthly*—was enrolled from 1925 to 1927. However, as demonstrated by its original name, "Mississippi Industrial Institute and College for the Education of White Girls," the school had nothing to offer Harrison.

Her restricted access to public education is of course on display in the nonstandard grammar and orthography of *My Great, Wide, Beautiful World*. Harrison had virtually no formal training as a writer. Yet the book's contents reveal how she seized opportunities to compensate for the lack, acts akin to perusing that magazine at work. *My Great, Wide, Beautiful World* shows her to be a patron of libraries and a pilferer of paper. Prior to embarking in Madras for Colombo, her last rupees went to lemon drops and books for the voyage. She conceived as "studying" her practice of poring over guidebooks and encyclopedias. She dabbled in Russian, engaged a German tutor, and spent hours discussing religion with Hindi, Buddhist, Jewish, and Muslim men. For more systematic instruction, during her itinerant years in North America, Morris relays, "wherever it was possible she attended classes at the Y.W.C.A. or night school" to "become an accomplished lady's maid" (ix); thereafter she "ambitiously took up the study of Spanish and French" (x). And of course, she wrote, copiously, as the existence of the book attests.

Morris is silent on the subject of Harrison's family. Their absence from the preface could be due to Harrison's own reticence, or it could reflect Morris's view of her work life as more germane. Also of note is that strong adjective in her characterization of the "sordid life" that "colored her early years" (ix), with its connotations of squalor and even vice. Did Morris's verdict of "sordid" stem from her assumptions about impoverished Black families, or was it prompted by revelations Harrison made in the course of their interviews?

None of her private letters even hint at such an upbringing. Harrison invokes her lack of privilege in a far more mediated way, aptly enough in discussing her working methods with Morris: "She said that I was a born Writer. I thought how far I was from what I was born" (FL2). The physical location of the exchange, the apartment in Paris where she and Morris worked on the book manuscript, was thousands of miles from Harrison's birthplace, and the social and psychological distance she had traveled from her unschooled, hardworking childhood was immeasurable. In dubbing her a "born writer," Morris alluded as much to her privation as to her innate talent, and Harrison revived the collocation to make it mean even more.

Juanita's mother died when she was six, at which time she was shifted to a different household. Although a passing reference proves that she knew her maternal grandmother, Lizzie Crigler does not seem to have raised her, nor did other family members take her in. Harrison informed Foster, "I went out from my dozon of Kin before I were old enough to know all their names" (FL3). The exact circumstances are left unspecified. However, in her interviews with Morris, she launched her oral history—the "true life story which I begain at six" (FL2)—at that pivotal leave-taking. The events of her life only invited narration after she was separated from home and family—that is, once she traveled. Startlingly, she cast the net result as positive, resulting in more agreeable conditions and all for the best. Having divulged that she "went out from" her family so young, Harrison continued, "[I] am sure I could never have trusted either of them my likes and dislikes from a tiny child were always so different." Her traveler's sensibility was forged from wary defensiveness at home.

Alleging she "could never have trusted" any of her family members is a strong charge, of a quite different order than acknowledging differing "likes and dislikes." In another letter Harrison expresses similar sentiment in explaining, "at 11 years I planned my life and in it I did not take neither man Woman nor Child. because I had looked at my kin and though they were kind in a way not one I was sure that I could never take for a true hearted friend not even the mother of my own mother so it was from that age that I took Jesus for my side by side every day friend" (FL2). When still a child she learned to rely, save Jesus, on no one.

The members of the household she joined after leaving her family were likewise less than supportive. She continues, "Then on with People that had known me from birth and they knew that I was a good clean obedient girl. and they were kind to me yet I knew that not one could I trust as a heart to heart friend altho. I tried many times and through something in some way it would be reveled to me that I could never trust them as a true heart felt friend" (FL2). Harrison repeatedly—near compulsively—reiterates her disappointment in family members and early associates.

The defensive assertion that she was "clean" also recurs in discussions of her past. In respect to her wish to publish under a pseudonym, in the same letter she insists, "not that there are any thing disgraceful because I have been careful from an early age to keep my life clean not for the public but for the honest feeling to know that I have not sin against Jesus, and in my own heart know that I have not disgraced myself nor any other human being" (FL2). She continues on to allude to a youthful scandal of some kind:

> Altho. I have been in the midst of Christian People I have never had one to give me a good straight forward advice that is one of the Christians weakest points gieveing Young Girls without mothers good advices. I am speaking from my own experence which I have had in nearly every state in the Union. and among my own Kin which I have many and also people that knew me from birth. They keep their mouths shut. and just as long as you hold out it remain Shut. but if you fail it open and never again close. and at the same time they had never giving you a Christians advice.

The allusion to failing to "hold out" conjures sexual temptation. Moral guidance withheld from a motherless girl resulted in bad decisions, a sullied reputation, and relentless gossip. Expanding the complaint, she then moves from dissatisfaction with her extended family to dissatisfaction with her racial community. "But its just in our race," she tells Foster, bitterly enough, "we will be kind up until we think that one count on us as the only friend then we let you fall. then after we feel dreadful about it. and we have a right to express our thoughts mine are from experence." The problem radiated far beyond her personal circle.

Given the vulnerability of Black girls in their employers' homes, "directly under white power within a system that condoned white male/black female relations," as David Katzman expresses it in his history of American domestic service, it is possible that the scandalous incident stemmed from employer coercion or assault.[42] If so, Harrison's insistence on her virtue reads as all the more poignant. Ayesha K. Hardison explains that "violating the codes of respectability remained particularly appalling for black women in the mid twentieth century," since such codes had long been "the primary defense against sexual violence."[43] While Harrison freely confessed to a "snapy eye for flirting" (309), she regularly stressed her core chastity.

Once she left Mississippi, the disheartening behavior of her home acquaintances found an afterlife in their inability to grasp the nature of her migratory quest or write the encouraging, reliable letters she needed. She kept in touch with them for many years, from at least as far afield as Canada and Cuba, but felt let down by their desultory correspondence. She recalled,

> Now when I was traveling in America in Canada and Cuba I tryed and they would wait a month or two and when they did write the first short Part would be excuses the next illness or deaths of ones that I knew nothing about and I am sure that they felt uneasy lest I would write and ask for a little money to get back People that are stay at home, are narrow minded so that they would stop answering after a while. (FL2)

Their expectation of her failure, of her likely need for money to slink home on, shows how much these "stay at home" people misjudged her aspiration and enterprise. At least in respect to their writing, however, they may deserve more credit. Their limited resources notwithstanding, they had internalized epistolary conventions, reporting on sickness and death and opening with apologetic explanations for late replies. However, lacking Harrison's literary abilities, they were unable to take the next step and break free of these same conventions, to be more creative and emotionally expressive in their letters.

Virtually all of Harrison's allusions to her childhood are negative. Perhaps hearkening back to her trauma as a neglected American child let down by "even the mother of my own mother" (FL2), she once identified a former employer as "the only American mother that I like" (292). Her early life was shot through with hardship and loss. Yet even so she took charge of her fate. "At 13 I begain to live it a little like I had planned" (FL2), she proudly stated. Incredibly, the execution of her life plan commenced while she was a girl in Mississippi, working as a maid, at the exact moment that the state was foreclosing on its African American residents.

Morris's portrait of the young Juanita dressed in "stiff" restrictive women's clothing—but shod in "laced bicycle boots" (ix)—is eloquent. While she has yet to come into her full mobility, she looks poised to leave. I posit that thirteen is when she began taking short trips to nearby attractions, funded by her own wages.

If so, she had ample company, as by the late nineteenth century, African American groups in the South routinely chartered trains for day outings, often to raise money for a church, club, or cause.[44] The practice upset white employers, discomfited by being deprived of their servants' labor, no matter how briefly. Linda K. Kerber explains that even as African American women were expected to always work, "the ideological legacies of slavery ... denied that what domestic laborers did was respectable 'work.'" As a consequence, they "found themselves trapped in an ideology of work riddled with a dizzying series of contradictions."[45] Their deliberate pursuit of leisure was thus deeply meaningful, as one way to cut through that "dizzying" belief system. Such pursuit was a keynote in Harrison's life. Her book repeatedly shows her desire for pleasure and change thwarting white people's desire to retain her services.

One early September morning, the train she was taking to Birmingham, Alabama, derailed, with scores of fatalities. *The New York Times* ran a front-page article about the tragedy:

> While rounding a curve on a high embankment near Berry, Ala., at 9:30 o'clock this morning the engine and four cars of an excursion train on the Southern railway leaped from the track and rolled over and over, smashing the coaches into kindling and causing the instant death of thirty persons and the injury of eighty-one others.
>
> With the exception of H.M. Dudley, train master of the Southern Railway living at Birmingham, J.W. Crook, engineer, and Roscoe Shelby of Columbus, Mississippi, all of the dead and injured are Negros who

had taken advantage of excursion rates from points in Mississippi to Birmingham....

There were ten cars to the excursion train, but the fourth broke loose from the fifth and with the heavy engine, plunged down the steep incline. The cars, which were packed with passengers, turned completely over several times and were crushed like eggshells, killing and crippling the inmates. Persons who have returned from the scene of the wreck say it is indescribable. The dead bodies of the Negroes were scattered in every direction and the moans and appeals for help from the wounded was heart rending.[46]

Even in its manifest pity, the sensational reporting exposes racial fault lines, naming only the three white victims while luridly conjuring "the dead bodies of the Negroes." In respect to the accident itself, that the train served working-class African Americans doubtless accounted for the lax safety standards. Published lists of the victims show that the majority were women and men from Columbus and adjacent towns. Harrison surely knew many of them. Yaeger suggests that the experience may have contributed to her decision to leave the South.[47]

Readers of *My Great, Wide, Beautiful World* learn of the Alabama accident in the course of Harrison's account of another catastrophic train wreck. On September 1, 1928, she was travelling from the city of Brno on a Paris–Budapest express, looking forward to taking the waters at Budapest's famous spas, when it collided head-on with another train. Twenty-one passengers were killed, and many others were severely injured. Since the stricken were concentrated in the first-class car directly behind the engine, they were of a much higher social class than the Columbus train wreck victims, including "several," according to the *New York Times*, "well known over a good part of Europe."[48] Harrison recalls in the book, "all of a sudden I was throwd across the compartment and hit my head. all the others jumped up and begain talking fast" (50). The ordeal prompts memories of the earlier accident: "I was in a Reck in 1903 on Sept. 1st when I was a little girl also a Monday but it was twice as bad about 100 killed and many wounded I didnt get hurt but many in the coch did so I knew just what to do" (53). Her date was off by a year, as the accident actually occurred on Monday, September 1, 1902. The more significant discrepancy is her belief that she was "a little girl" when she was in fact close to fifteen. Her shock and vulnerability likely contributed to the memory of being very young. That she knew "just what to do" all those years later suggests the impact of the disaster.

Recounting the later accident cuts through her usual silence to provoke the only childhood scene to appear in the book—a near emblem of thwarted mobility.

At the same time, her taking such a trip at all speaks to her agency. Her presence on that train gives us a sense of how routine travel had already become for her and the extent of her self-sufficiency. Despite her youth she appears to have been traveling on her own. Birmingham, a seventy-mile journey east across the Alabama state line, was one of the most accessible destinations from Columbus. Much larger than Harrison's hometown, this growing industrial center offered visitors a foretaste of the new South. Plans were already underway for its iconic monument, the fifty-ton cast-iron "Vulcan Statue." Black excursionists would have gravitated to the African American business and social hub along Fourth Avenue, with its many restaurants and entertainment venues, and perhaps toured the neighborhood of Smithfield, which boasted gracious homes owned by the city's African American middle class. Hopefully Harrison later made a happier journey there. Birmingham helped form her geography of resistance, and as the next chapter discusses, it may well have been the first station of her migratory circuit.

It is intriguing to think about where else she might have gone, crossing state lines, before leaving Mississippi for good. Her home state is unusual for having both north–south and east–west train lines, expanding its residents' travel options. The 1904 St. Louis World's Fair was held in St. Louis, Missouri, when Harrison was sixteen. Considering it was more than four hundred miles to the north, she could scarcely have attended, but she must have dreamed of doing so—undeterred by reports of the discrimination that African American fairgoers faced. Like its Chicago predecessor, the St. Louis World's Fair was organized around anthropological exhibits devoted to different nations and ethnic groups. Its model villages displayed natives of Central Africa, Sri Lanka, the Arctic, and the Philippines, the latter of particular interest due to its new status, following the Spanish-American war, as the United States' first colony. Closer to home, it also offered visitors a replica of "the Old Plantation," complete with slave cabins, which had become a staple of American expositions. Harrison loved such grand-scale expositions, their insidious colonial and racial premises notwithstanding. Later, even as she was actually travelling the world, she planned her itineraries to maximize her time at "world's fairs."

Harrison spent just over a decade working as a domestic servant in Mississippi. Her introduction to service took the form of "an endless round of cooking, washing and ironing" (ix), in Morris's words. The long hours were doubtless compensated with the most meager of wages, boosted, according to Southern custom, with leftover food and other household remnants. As she grew older, her positions would have grown more specialized, since the abundance of domestic workers in the South led to the distribution of "highly specialized work tasks" among women variously employed as cook, laundry worker, nurse, housekeeper, and maid.[49] Initiated so young, Harrison acquired a formidable set of domestic skills, which she later fortified with vocational courses.

Her book shows that she identified primarily as a housekeeper, responsible for the affairs of a household. Next in order were perhaps posts as personal maid or children's nurse. Willing to do laundry on occasion, she was least likely to accept work as a cook. Standard for servants in the South, her early positions would have given her considerable discretion in executing her duties, which means that later in life she had to adapt to closer supervision elsewhere.[50] Given Harrison's certainty about the superiority of her ways, she surely chafed at the difference. "One thing She doesnt go about the house nagging you" (246) she allowed of an abusive employer in Paris.

At the outset of her world journey, prospective employers in London were so importunate that she happily concluded, "That means I can get a place easyer here than in the U.S." (14). Her nationality, skill set, language abilities, and, not least, charisma meant she was always in demand. There were simply no other workers like her passing through Europe and Asia. Thus she could often dictate the conditions of her employment.

During her years of service in Mississippi, in high contrast, she would have had virtually no latitude for setting her own terms. Prior to World War I, the glut of labor in the South left servants with no "bargaining power."[51] Aside from some jobs at tobacco plants, domestic work was the sole occupation open to African American women, whose numbers were continually replenished by young girls consigned to service.[52] Abroad, Harrison was singular; at home, she was typical. Moreover, the South's rigid caste system checked African American workers just as much as did economic forces. A society structured by racial hierarchies could not countenance, to the slightest degree, servants stipulating how they would work.

The testimony of "a Negro nurse" that ran in a 1912 issue of *The Independent* suggests the nature of Harrison's employment in Mississippi and what she might have continued to experience had she stayed. "More Slavery at the South" was compiled from interviews with a veteran domestic worker in Georgia. While the language is the reporter's and includes some evident editorializing, the conditions relayed have the stamp of firsthand experience. The nurse was initiated into service at the same age as Harrison. She recounts,

> For more than thirty years—or since I was ten years old—I have been a servant in one capacity or another in white families.... I was at first what might be called a "house-girl," or, better, a "house-boy." I used to answer the doorbell, sweep the yard, go on errands and do odd jobs. Later on I became a chambermaid and performed the usual duties of such a servant in a home. Still later I was graduated into a cook.... During the last ten years I have been a nurse.[53]

In her current position, she continues, "I am not permitted to rest. It's 'Mammy, do this,' or 'Mammy, do that,' or 'Mammy, do the other,' from my mistress, all the time."[54] She never rose above subsistence wages, and the social indignity sharpened the economic injustice; even the children would not address her as "Mrs."

As the nurse suggests, servants' relationships with their female employers could be particularly fraught, brought into an unequal intimacy that was "so much a function of white dominance and black subordination." The words here are Katzman's, who concludes, "The only effective way most blacks had to counter the power of Southern white mistresses was to leave the South."[55] Harrison, of course, did just that, and she went on to form often rewarding, and always negotiable, relationships with a number of white women employers. In Mississippi in 1930, can one imagine the servant-employer interaction she recorded in Paris that year? "I like my job ever so much," she divulges. "If we had a big scrap when I was ready to go home in the evening the Lady gave me 10 francs to by icecream if it was a little scrap I got only 5 francs. Now we are happy and I massage her head some mornings and get the francs just the same" (251–52). Her book documents many minor power struggles between herself and the women for whom she worked, as in this instance usually humorously narrated. Yet her very commitment to recording them—and to showing herself to be the victor—hearkens back to a time when she didn't have such license. As conjured by Morris in its preface, the primal scene of *My Great, Wide, Beautiful World* is the "overburdened household" (ix) in Mississippi where the young Juanita toiled. This scene is displaced by Harrison's staging of numerous domestic workplaces around the world, whose duties she carried very lightly.

A later chapter will address the shifting national and ethnic claims she makes in the book. For now it suffices to state that within it she acknowledges no ties to her racial community of origin. While in her private letter to Foster Harrison complains about "our race," *My Great, Wide, Beautiful World* includes this oddly distancing aside: "Now no body can cook cabbage to beat the Irish of Cork not even the American Colored Southerners" (8). We cannot know whether her seeming detachment stemmed from a disinclination to call attention to her regional and racial identity in a narrative largely directed to white readers in other parts of the country and overseas or if it stemmed from a genuine sense of estrangement, from not perceiving herself to be an "American Colored Southerner." What is certain is that she left Mississippi well behind.

It is easy simply to celebrate Harrison's freewheeling ways, unbound to people, places, or jobs. This is the prevailing critical stance. Profiles admiringly quote statements such as that she made on boarding her ship in Hoboken, "I was happy that I had no one to cry for me" (1). By staying single and avoiding travel companions, she was able to stay on the move and focus on her own needs. By choosing not to commit to any employers long term, she could do the same. In her book, she regularly mourns the places she leaves behind but never mourns the people she came to know there. However, this lack of attachment to other human beings—or perhaps, her efforts to resist such attachment—is so consistent as to appear as much a symptom of trauma as a sign of self-determination. Her book shows her habitually leaving other people just as the bonds between them strengthen, and she made a point to keep her distance even from cherished friends.

We might posit early family rupture as the root cause. Losing her mother as a small child, afterward Harrison felt little real kinship with her many relatives. Withholding the support and counsel she needed, they let her down. The scandal she endured strengthened her Christian purpose, but the church, too, she concludes, betrayed her. These experiences made her self-reliant, forging the core sensibility that led her to forego intimate human connections. They also made her a writer, in that voluble correspondence became her preferred method for managing far-flung, attenuated personal relationships. In 1935, as the close of her world journey hove in sight, she remarked, "Life wouldnt be worth while looking at the same faces day after day. I trust I'll never endue such a sad life as that" (307). She trusted right—she did not.

CHAPTER 2

Vagabond Housemaid

Harrison never looked back after leaving her Mississippi hometown. There is no record of her visiting the state again, much less living there. Nor do her writings exhibit a mote of nostalgia for the people, places, or events of her youth. Instead, her fond memories center on the years following her departure, when she roamed the country as a "Vagabond Housemaid," as her admirers came to call her.[1] She recollected that seventeen-year period with much satisfaction: "I did what I wanted saved what I wanted and I traveled about from State to State City to City in the Union Canada and Cuba" (FL2). Proving that her North America wanderings were as studded with adventure as her time overseas, her editor in Paris was riveted by her reminiscences. If only they had been preserved.

This chapter charts those itinerant years, 1910 to 1927, from when Harrison left Mississippi until her world journey commenced (the next chapter delves deeper into her mid-1920s interlude in Los Angeles). The episodic record of her domestic migrations suggests that as a young woman she embraced nomadism both to control the conditions of her labor and to satisfy her craving for novelty. Her transience also spurred her literary production: For Harrison, authorship was always intertwined with mobility. We know that she was writing letters during those years, given her complaint about the dilatory, downbeat replies she received in their course from "stay at home" Mississippians (FL2); considering her later practice, she most likely wrote to some of her former employers, too. Unfortunately, however, the collective archive includes no extant texts in her hand from that period. Nor are there any public records that shed light on it, save the 1950 passport application she submitted to the US consulate in Buenos Aires.

Supplementing that crucial resource, one way to recover Harrison's experience is to extrapolate from that of other African American migrants and domestic workers. Another is to make a scrupulous reading of *My Great, Wide, Beautiful World*. For one, it includes many allusions to past events in North America. In addition, its inventory of Harrison's activities overseas suggests the nature of the domestic ones that preceded them. The book encourages a process of back formation, of mapping the patterns of work, leisure, travel, and writing it displays onto this earlier era.

Just as important as the biographical insight it yields, reading *My Great, Wide, Beautiful World* in this way helps conceptualize the text itself, as a Great Migration narrative that gives further expression to the unprecedented demographic shift in which close to a tenth of the nation's African American population left the South. Lawrence Rodgers has identified Dorothy West's *The Living Is Easy* as the only female-authored work to represent the movement, but Harrison's book gives it some company.[2] Although set in European and Asian locations rather than North American cities, it too portrays a Southern protagonist who leaves her homeland to pursue a more gratifying life elsewhere. Driven by Harrison's Southern past, this book about international travel contains a shadow narrative of American migration.

The previous chapter discussed how the life program Harrison drafted at age eleven gave material form to the "bright vision of templed cities" (ix) she happened across in a magazine, which inspired her to become a world traveler and perhaps a travel writer as well. Shortly thereafter, initiating her life-long practice, she began taking short outings such as the calamitous rail journey to Birmingham. "At 21" she "had it just right," which is when she left her home state to begin crisscrossing North America. She identified that year as her life's hinge, reporting, "I have lived just as I do now from 1910" (FL3). (The dates do not quite add up, as Harrison was twenty-two for most of 1910, but the math is close enough to persuade me.) Living "just as I do now" evokes her free self-sufficiency. Unencumbered by property, possessions, or family, she moved from job to job and place to place, stockpiling her wages for the next trip. She was not, however, drifting. She mapped out both her travel itineraries and her life course with much deliberation.

Shortly before Harrison left Mississippi, the state's most salient migration was westward to Indian Territory. In preparation for Oklahoma statehood, the federal government began allotting individual land grants to tribal members, prompting thousands to rush to Indian Territory before tribal rolls closed. Applicants included not only Choctaws and Choctaw freedmen with blood or cultural ties but also many white, Black, and Creole people who hoped to bluff their way in.

Despite her purported Native American ancestry and despite the fact that a branch of her family emigrated to Oklahoma in the 1890s, Harrison did not make such a bid. Instead, her passage through Midwestern and Northeastern urban centers puts her in the vanguard of a much larger movement. In 1916, the Pennsylvania Railroad Company set a precedent in recruiting Black Southern workers, and a plethora of new opportunities followed. Close to four hundred thousand African Americans are estimated to have left the South between 1916 and 1918 alone. By the 1930s, the number was more than 1.5 million.[3] James Weldon Johnson recounts, "Migrants came north in thousands, tens of thousands, hundreds of thousands—from the docks of Norfolk, Savannah, Jacksonville, Tampa, Mobile, New Orleans, and Galveston: from the cotton fields of Mississippi, and the coal mines and steel-mills of Alabama and Tennessee; from workshops and wash-tubs and brickyards and kitchens they came."[4] While this movement is often associated with men seeking industrial jobs, many women migrated too—acknowledged here by the nod to "wash-tubs" and "kitchens"— even as they were mostly restricted to domestic service.

As usual, with her 1910 exit from Mississippi Harrison was in the forefront, just ahead of the migratory wave that would soon wash over the nation. World War I drastically curtailed European immigration to the United States even as it necessitated increased arms production, leaving Northern and Eastern factories with a labor shortage and opening up jobs to Black citizens. In 1915, over one million Europeans entered the United States; a year later, scarcely three hundred thousand did.[5]

During this same period, conditions in the South were deteriorating, spurring its residents to leave. Harrison's birthplace, Lowndes County, lost 10 percent of its African American population, and Mississippi was in crisis statewide. Its cotton industry had been decimated by a boll weevil epidemic, resulting in fewer jobs in the short term and crop diversification that dampened labor needs in the long term. Due to both economic and political factors, moreover, many Black farmers lost the land they had acquired in Reconstruction, forcing them into sharecropping. These reversals coincided with increased anti-Black violence. In *Wayward Lives, Beautiful Experiments*, Hartman foregrounds emotional forces, too: "*Something else* was never listed as one of the reasons people left home, not only the appalling and the verifiable: the boll weevil, lynching, the white mob, the chain gang, rape, servitude, debt peonage; yet the inchoate, what you wanted but couldn't name, the resolute, stubborn desire for an elsewhere and an otherwise."[6]

In some respects Harrison was representative, a Black woman leaving the South for greater scope. Yet as Arna Bontemps and Jack Conroy remind us in their 1945 study of the Great Migration, *Anyplace but Here*, each migrant was "motivated by a set of circumstances peculiar to himself."⁷ Moreover, Harrison's mobility was of extraordinary scale. Virtually none of her peers moved ceaselessly over more than a score of years through more than a score of countries. She was motivated not solely by the punishing environment of her Mississippi youth but also by her private urges and unique psychology.

"Solving the puzzle of motivation is the key," Susan Armitage has stated in respect to successful biography.⁸ What inner forces compelled a subject to make the choices she did? Perhaps Harrison's bottomless need for travel was innate. Perhaps it developed after she "went out" from her family as a child (FL2). Without minimizing the systemic oppression that drove the Great Migration, I posit family trauma as the wellspring of her perpetual departures. We might recollect Richard Wright's characterization of his escape from Mississippi in the conclusion of his memoir, *Black Boy*: "With ever watchful eyes and bearing scars, visible and invisible, I headed North."⁹ Yet this reasoning risks sounding too dark. Every page of Harrison's book attests to the joy she took in immersing herself in a stream of cities, countries, and landscapes.

In February 1950, as her long residence in Buenos Aires neared its end, she applied for a new passport in anticipation of "traveling by rail" through Bolivia, for the "purpose of pleasure." By then, she could identify her profession as that of "Author"; she also stated that she had "never married." The requirement to record her previous residences resulted in a skeleton chronology of her North American movements. In filling out the form, Harrison stated that between 1887 and 1917 she lived "continuously in the United States" in Mississippi, Alabama, Michigan, Illinois, New Jersey, Pennsylvania, New York, and Florida, listing the states in that order. From 1917 to 1922, she was outside the country. Between 1922 and 1925 she was in Louisiana, Texas, Kansas, and Indiana, and then in New York once again. From 1927 to 1935—the span of her world journey—it was "Europe and Asia." The application also documents her residence in the "Hawaiian Islands," 1935 to 1939, and her 1940 return to the United States.¹⁰

The record is incomplete, however. It does not disclose that she was in Canada and Cuba during the first long period outside the United States; identify her American locations between 1925 and 1927; or include her known residencies in Iowa, Colorado, and California. Absent, too, is an earlier sojourn in Hawaii, the six months she spent in Honolulu in 1925.

She may well have lived in even more states, especially considering her claim, made as early as 1931, to have been in "nearly every state in the Union" (FL2). In 1950 she was sixty-two, with four decades of travel and itinerant work behind her. In completing the paperwork, she overlooked or even forgot some of her past homes, now literally uncountable. Or perhaps, having maxed out the form's allotted lines for the information, even wedging in additional text between them, she resorted to picking and choosing.

Nevertheless, we can surmise that she listed her American states of residence in roughly chronological order, given that the sequence starts in her native state of Mississippi and ends in New York, her last US location before crossing the Atlantic. Thus her trajectory can be mapped: from Columbus in eastern Mississippi to Alabama just across the state line, north to Illinois and Michigan, east to Pennsylvania, New Jersey, and New York, and then south along the eastern seaboard to Florida. Florida was her last American home before she went to Cuba, which brings her to 1917, approximately seven years of roving. Prior to 1922 she also lived in Canada, on both the west and the east coast. The three years that followed include time again in the South—Texas and Louisiana—and the Midwest—Kansas and Indiana. She lived in Los Angeles in the mid-1920s, and she launched her world journey in 1927 from New York.

Alighting in Alabama was a typical first step for Mississippi migrants, who like other Southerners often made "a series of stops" within the South before continuing on to the North or the West.[11] The odds are high that Harrison started out in Birmingham, a "concentration point" for Southern migrants.[12] Only seventy miles from Columbus, Birmingham was already within her ken, as we know from her aborted attempt to go there as a girl. This rapidly industrializing city offered higher wages than did neighboring Mississippi, and it also served as a center to learn about even better opportunities elsewhere. Sociologist Charles S. Johnson's record of a 1917 interview suggests its reputation:

> Man a laundryman in South earning 9 dollars per week. . . . [His mother] had occasion to go to Birmingham. There the people were leaving in large numbers for the North, mostly men. She asked why. They said, higher wages. Yet she knew that wages in Birmingham were twice as high as in Meridian [Mississippi]. The people in her home town had been approached by agents but doubted. She herself could not believe. Went home and told her son of Birmingham and urged him to go and see for himself. He left in December, in 3 weeks he wrote home. "Everything is just like they say, if not better."[13]

Johnson captures a characteristic swirl of news, rumors, and reports. The initial moves out from home communities were facilitated by networks of information and aid. New migrants wrote to those still at home, urging them to find out for themselves, and their letters could include offers of places to stay or even money to travel on. Did Harrison benefit from this kind of support? She never states that her early relocation was embedded in such a network, but this does not prove that it wasn't. Indicating they were accustomed to assisting those on the move, she noted that home acquaintances in Mississippi were worried that her letters had ulterior motives: "I am sure that they felt uneasy lest I would write and ask for a little money to get back" (FL2). She may even have left Columbus with other people. At the least, while still in Mississippi she must have heard or read news from migrants describing rewarding experiences in other places.

From Alabama she likely went next to Chicago. The north–south lines of the Illinois Central Railroad made the city "the logical destination" for those leaving Mississippi, Arkansas, Alabama, Louisiana, and Texas, while the *Chicago Defender*, widely distributed by railroad porters throughout the South, vigorously promoted migration and led Southerners to associate Chicago with "the promise of the North."[14] The 1893 Chicago World's Fair further contributed to the city's allure.[15] In his landmark study, *Land of Hope: Chicago, Black Southerners, and the Great Migration*, James Grossman states that the fame of Harlem notwithstanding, "It was Chicago that captured the attention and imagination of restless black Americans."[16] More materially, Chicago had more factory jobs than New York, whose economy was driven less by industry than by service.[17] Despite the severe housing discrimination they endured, up to seventy thousand African American migrants settled in Chicago between 1916 and 1919 alone, and it was also an important way station for thousands who later moved on.[18] Contemporary memoirist Morgan Jerkins remarks, "If your family was from Mississippi, chances are they wound up in Chicago."[19]

The opening lines of Langston Hughes's "One-Way Ticket" well portray a migrant's resolve:

> I pick up my life
> And take it with me
> And I put it down in
> Chicago, Detroit,
> Buffalo, Scranton,
> Any place that is North and East—
> And not Dixie.[20]

Note that Chicago is the first stop in "One-Way Ticket," while the second one is Detroit, another major destination for Southerners. A passing reference in *My Great, Wide, Beautiful World* indicates that the latter was likely Harrison's place of residence in Michigan: she was careful to note when she met people with connections to cities where she herself had lived, and she comments of a woman she spoke with in Nazareth, "Her Husband and boy living in Detroit and Her Mother with a brother in Havana so I felt quite near them" (71). As early as 1910, Detroit had a population of almost half a million immigrants from Lebanon, Syria, and Yemen, who like the African American newcomers were drawn by its many jobs in industry, especially at Henry Ford's automobile plant. Perhaps their presence increased Detroit's appeal for her.

Upon moving North, like other migrants, Harrison could have received aid from the many organizations newly established for their support. Some were dedicated to African Americans, such as the Chicago Urban League or the nation's network of Phyllis Wheatley clubs, which took their name from the eighteenth-century poet. Yet despite their social services, economic opportunities, and cultural depth, her stays in these Midwestern cities were brief, as within three years of leaving Mississippi she had reached New York City, probably by way of Pennsylvania and New Jersey. Unusually for her domestic migrations, we know the very year she was there, thanks to her comment about attending a lecture at "Cargine [Carnegie] Hall in N.Y.C. in 1913" (26). This makes for at most three years from Mississippi to Manhattan—far from a beeline but still rather swift, with Harrison settled in New York well before the 1920s mass influx of African Americans.

The subject of the Carnegie Hall lecture was John Milton's epic poem, *Paradise Lost*. Comparable to her diligent language study and enrollment in night school courses, Harrison's attendance at this quintessentially middlebrow event is a telling sign of her aspiration. Brimming with intellectual confidence, undaunted by the learned lecturer or lofty venue, she rejects the interpretation she hears and the poem's very premise: "I do not agree and have ofen felt sorry for the speaker as it was on Paradise Lost. its not lost its right here on earth and I have seen so much of it" (26). Beyond pinpointing her physical location, this account sheds light on her subjective location, too.

Like Honolulu, Paris, Nice, Los Angeles, and Buenos Aires, New York was one of the cities to which Harrison periodically returned. She was in New York for at least three separate intervals. Her book refers to residencies in 1913 and 1927, and the passport application in Buenos Aires records that she was also there sometime between 1922 and 1925. Perhaps her recurring New York stints are not unrelated to the subject of her disclosure, "For years in N.Y. I was a love of the Army" (102). This confidence, along with constituting her sole known allusion to an enduring romance, proves that New York was one of her longer-term homes, with her residence measured in not months but years. Given her proclivity for urban places and verified presence in Manhattan, "N.Y." almost certainly refers to the city, not the state. Perhaps the soldier was stationed at Fort Hamilton in southwestern Brooklyn.

Directed by Morris's preface, most profiles of Harrison identify Los Angeles as her last US residence prior to her world voyage. In fact it was the second to last. She sailed from the east coast, embarking in Hoboken, and she identified her summer 1927 position in London as "my first job since I left the swell apt on West End Ave N. Y." (5). The West End was an emerging upper-class neighborhood in midtown Manhattan, adjacent to the Hudson River. Now extolled for its commodious residential buildings, "its classic sixes and sevens, sprawling apartments with two or three bedrooms and a maid's room with bath," it made a suitable employment district for this veteran housekeeper at the apex of her American career.[21] The job sounds similar in prestige to the one with Myra and George Dickinson in Los Angeles that she had just left, and to her future positions with influential people in Europe.

This final New York period is rather hard to account for. Indicating she had stored her possessions in California in anticipation of an extended journey, when questioned at Japanese immigration, "Where are your baggage?" Harrison replied, "I left it in Los Angeles 7 years ago" (285). She might have taken the West End job to accrue extra funds for the trip. Yet if she needed more money to travel on, why not just continue working for the Dickinsons in Los Angeles? Maybe they were out of town. The couple regularly took long vacations, often in spring, in far-flung places like France and Hawaii. If they had done so that season, they would have had no need of Harrison's services, perhaps prompting her to take a stopgap job in New York until her June departure.

It is also possible that the deciding impetus for the transatlantic voyage only came during her time in New York. The Hoboken port just across the Hudson, from which German steamers plied the Atlantic, was temptingly near. Or to look for a less literally concrete inspiration, Harrison may have been inspired by the international currents that were circulating in New York, especially among African American intellectuals and artists. By the late 1920s, the Harlem Renaissance was accelerating toward its peak, with the publication of key works that include Alain Locke's anthology *The New Negro*, Langston Hughes's poetry volume *The Weary Blues*, Jean Toomer's genre-defying *Cane*, Jessie Redmon Fauset's novel *There Is Confusion*, and the single issue of the magazine *Fire!!*

The movement had a conspicuous transnational dimension. Many of its participants spent time in the Caribbean, Mexico, and Europe, especially France and the Soviet Union. New York City was the American corridor through which numerous Black women and men passed on their way to other nations, and Harrison's choices resemble those of Renaissance personages including Fauset, Hughes, Gwendolyn Bennett, Countee Cullen, Zora Neale Hurston, and Wallace Thurman. She left New York just before the publication of Nella Larsen's *Quicksand*, which tracks the migratory experiences of its protagonist, Helga Crane, including intervals in Alabama, Chicago, and Harlem and another in Copenhagen modeled on Larsen's own. Living in New York may have exposed Harrison to her compatriots' reports of their experiences abroad, including the relative freedom from racial prejudice they found.

Statements abound about the impact of New York City on new arrivals. "The dreams of my childhood and the purposes of my manhood were now fulfilled," Frederick Douglass declared in respect to reaching New York after his escape from slavery, and in the twentieth century, exultant newcomers reprised his sentiment as they encountered a vibrant Black community in the North.[22] Hughes's recollection in his autobiography, *The Big Sea*, is exemplary: "I came out on the platform with two heavy bags and looked around. It was still early morning and people were going to work. Hundreds of colored people! I wanted to shake hands with them, speak with them. . . . I stood there, dropped my bags, took a deep breath and felt happy again."[23] To many African American transplants, New York offered liberating, novel forms of racial community.

Harrison, however, did not voice this manner of enthusiasm about New York or suggest that it was transformative. While she certainly liked the city, her references are occasional and low energy, as when she describes an Irish market as resembling "the low East side in New York only smaller" (8) or observes that the "beautiful View" from her employers' hotel suite in Madrid "look like Central Park in New York" (211). For Harrison, New York appears to have been one important place among many others, and far from the most so—Havana, Nice, Los Angeles, and Honolulu have that honor. Still, she intimates that she saw living there as a totem of modernity and privilege. As ironic contrast to her bedraggled appearance in Edinburgh, she wryly states, "I left New York with a ten dollar hat on my head" (11), figuring the city as a locus of sophistication and abundance. In a third-class rail carriage en route to Mumbai, she likewise comments, "In my red shawl and black vail on my head no one would think I had seen Broadway" (112). Registering the impressions of a working-class Black woman in New York, even these desultory remarks are valuable as a counterpart to the oft-quoted affirmations of her male contemporaries.

The city's impact on Harrison was likely delivered not by its Black arts movement per se, but rather by its highly mobile population and the attendant multicultural exposure. After all, Rodgers points out, "It was Harlem's migratory aspect that predicated the existence of a renaissance sensibility."[24] Already by 1910, over three-quarters of New York's Black population had been born outside the state.[25] Locke effused over the social energy generated by "the first concentration in history of so many diverse elements of Negro life." New York, he declared, "has attracted the African, the West Indian, the Negro American; has brought together the Negro of the North and the Negro of the South; the man from the city and the man from the town and village."[26] While Locke does not acknowledge them, women were there too, and Harrison surely felt at home among this diverse group.

She would later attract notice from major Harlem Renaissance players. "I LOVE her," Carl Van Vechten announced in urging the *Atlantic Monthly* prequel on Hughes, and he gave Gertrude Stein the same counsel. Hughes went on to pronounce *My Great, Wide, Beautiful World* "swell" and recommend it to friends.[27] Langston Hughes read Harrison! Locke read her too, although he dismissed her as a "moronic menial."[28] Era Bell Thompson, the long-time *Ebony* magazine editor and foreign correspondent, acknowledged her influence in her 1946 memoir. "Ever since reading Juanita Harrison's Great Big Beautiful World," she recalled, "I had been threatening to go around the world, too."[29]

After New York, Harrison went south to Florida. She does not specify where she lived, but her book does reveal that she at least passed through St. Augustine. Florida was the obvious point of departure for Cuba, with regular steamboat service to Havana from coastal cities such as Tampa and Miami. Harrison was probably first in Cuba around 1917, when the island was on the verge of becoming a major American tourist destination. She was still quite young, scarcely thirty. With its coupling of the rustic and the cosmopolitan, Cuba made an abiding impression, and I will save a discussion of it for a later chapter, as a pivotal location in her Latin trajectory.

Until this juncture, her path displays some evident spatial logic. Thereafter it is harder to account for. Her residencies in Cuba and Canada were contiguous. Yet from Havana to Ontario alone is a fourteen-hundred-mile journey, and we know that the Canada interval included time in Vancouver, British Columbia.[30] Next she was in Louisiana and Texas—far to the west of Cuba and far to the south of Canada—before moving north to Kansas and Indiana. This stage of her peregrination does not fit into any neat schema. Even shelving questions of geography, it is hard to reconcile socially, economically, or psychologically. Harrison's chain of removals led her toward neither more liberal territory nor greater urban opportunity, or even toward replenished family and community ties. She did not return to her Mississippi hometown, Columbus, but she did return to her home region, the South, with its systemic racism and abysmal service conditions. That none of her extant texts give any detail about these additional periods in the South contributes to the mystery. Her life path denies readers the "symbolic geography" identified as characteristic of African American literature in Robert Stepto's classic study, whether the "narrative of ascent" tracking the protagonist's journey north or the "narrative of immersion" predicated on a return journey to the South.[31]

Even as we investigate Harrison's motives, we would be well advised to heed her pronouncement: "We all have fancys." She elaborated, "I did not like the People of Madrid but adored the Sevillians and so it was in the States of the Union in some I liked the People and some I didn't" (FL2). Her US destinations of choice might have been determined as much by her personal "fancys" as by socioeconomic factors.

One could wish she had offered more comment about her views on the different states and their "People." Ironically enough, considering it became one of her base locations, her only confirmed dislike was southern California. Just as they depart from the dominant geographies of African American discourse, Harrison's looping itineraries disrupt westering mythologies. We never see her dream about making a new start in the West or hear her echo her editor's identification of her move to Los Angeles as a "happy turning point" (x) in her life. Nonetheless, in the 1920s she began to shift her American axis westward, living in Denver, Los Angeles, and Honolulu. These cities were gateways to grand landscapes and bountiful outdoor leisure, and they also supported the multicultural communities she preferred.

We don't know much about the actual positions Harrison held during her nearly two decades crisscrossing North America. While her book includes fond references to leisure activities in the United States, it reveals precious little about her work. There is only this single, intriguing vignette: "I knew the English servants very well I worked in a big house in Iowa with them and they were selfish and jelious I have often hear it that they were the heardest of all servants to get along with I also knew the English ladies I worked for one in

Canada and one in Cuba. They have their servant problems here" (11–12). How did she come to inhabit a fractious English "big house" in Iowa? The revelation invites us to consider whether she was as attracted to "foreign" households in the United States as she was overseas, gravitating to employers who were not native citizens of the countries in which they lived.

Her report of open discontent in the Iowa household intimates a clash of cultures, between American and English forms of service. The arduous initiation Harrison endured in her youth may have contributed to her steady employment across the United States. "Invisible and silent," African American servants in the South had to suppress their personalities and opinions, and the wary habits Harrison formed during the first stage of her working life could have helped her get and keep jobs elsewhere.[32] Relatedly, her service jobs may have supported a traveler's sensibility in that they conditioned her to moving between different milieus and fostered the same social skills that the adroit wayfarer plies. Laboring throughout her youth in private homes in the South, Harrison was in direct contact with white people in power. Historians have shown that due to such contact, African American domestic workers "had a mobility denied to others of their race," both social and spatial.[33]

Harrison's peripatetic habits do not seem to have diminished her desirability. Indeed, by rendering her labor a scarce commodity, they may even have heightened it. Again she was remarkable only in degree, not kind. Katzman observes of the era's servants that "as a group they were movers," spurred onward by poor service conditions and heartened by a nationwide labor shortage. In addition to the arduous duties, servants across the United States complained about the dearth of privacy, denial of social status, and deluge of loneliness endemic to the profession.[34] "Constant change"—the most reliable way to improve the terms of their employment—became their "hallmark."[35]

Harrison's incessant travel is sometimes read as the ability of a working-class woman to seize upper-class experiences. The ubiquity of leisure tourism today can make one lose sight of the fact that historically, the poor were the most itinerant. The lack of real estate or other property makes mobility both more fruitful and more imperative. Harrison's resolute transience only amplified standard working-class practice. At the same time, it was likely further motivated by racial factors. African American women and men associated immobility with antebellum slavery, and the converse association, between movement and independence, continued well into the twentieth century.[36] Elizabeth Stordeur Pryor argues that since they were "core features of American citizenship," Black activists had long targeted "access to transportation, the processes of travel, and indeed mobility itself."[37] Every time she moved on to still another city, state, or country, Harrison proved her freedom.

In the latest extant letter we have from her, written in the mid-1940s, Harrison protests her consignment, as an African American, to a life of service, deploring the fact that "we are the foot ... stool, of every one. Because we serve every thing and every body."[38] Her complaint about being relegated to a service caste makes no reference to gender, but as a woman she was even more constrained. As late as 1970—three years after her death—the majority of wage-earning African American women were employed as domestic workers.[39] Despite Harrison's ability to thrive in adversity, to be restricted to an occupation whose fundamental premise was a denial of personal autonomy must have been harrowing.

During her years working in the United States, she strove to advance within the restricted field of domestic service by professionalizing her skills. Her *Atlantic Monthly* editor, Ellery Sedgwick, maintained, in his usual embellished, patronizing style,

> All day she trained herself assiduously to be a ladies' maid; in the evening she attended night classes at the Y.W.C.A. and picked up enough French, Italian, and even Spanish to make her simple wants readily understood. At Los Angeles her extraordinary competence endeared her to her employers. She remained four years in their service, and she graduated with the perfected knowledge of the art of arts—how to make people comfortable. She could cook, wait, valet and keep house.[40]

(When I read Sedgwick's evaluation of her language skills as adequate to communicate "simple wants," I always remember how he was roundly mocked for his abysmal Spanish.) Despite the distracting hyperbole, his comments capture Harrison's enterprise. She herself portrayed herself as a modern housekeeper who implemented efficient new methods. In describing her work at a home for disabled children in Barcelona, for example, she boasts, "I have caused some changes the Family eat early and quicker and I did not like the pot the soup was served out of so a nice new one came today. they are amused at me I get through with my work so quick and it is done so well" (175–76). She was fast, clean, and unfettered by convention.

Following her *Atlantic Monthly* debut, two of Harrison's former employers published letters about her in the magazine's Contributors' Column. Their patronizing tone notwithstanding, they too indicate how her personality coupled with her work ethic made for a highly attractive worker. Helen Rose of Cornwall, her long-time employer in southern France, disclosed, "We were very fond of her and found her an amazing character, besides being a very good servant."[41] From Los Angeles, George Dickinson reported, "She always made friends with whomever she worked for and a great many gifts were given to her."[42] Such statements echo Morris's claim that "her employers invariably became her friends and raised her salary in the vain hope of keeping so excellent a servant" (x). While any identification of Harrison's employers as her friends must of course be read skeptically, they do suggest more than usually warm relationships, as well as their rewarding financial outcomes.

Harrison gravitated to relaxed, amiable workplaces, her book shows. Her description of a Valencia household is characteristic: "They have 3 other Girls I only do the work I like to do the other do the rest.... the Lady say I must not ware a hat but I did and she could not get over it" (190). She also notes, "I wouldnt sleep in as I had planned to spend just two weeks here and I wanted to be free." We can surmise that in the United States, too, to boost her freedom she often opted not to "sleep in," given not only her voiced preference but also prevailing custom in her native South. Southern servants usually maintained their own households, finding the greater privacy well worth the additional expense.

Exacting about her terms, Harrison suggests that she was usually able to satisfy them. She preferred urban over rural households. She valued employers who provided "good food" and "a good bed" (183) and were liberal with presents and bonuses. Suggesting hard lessons learned, she avoided jobs that entailed taking care of young girls or pets, sharing a bedroom with a woman employer, or working for widowed fathers.

Certainly her wishes were not always granted, but when the negative outweighed the positive she was sanguine about moving on. Only rarely did she have to bear with a job she did not like to fund the next stage of her travels. She could quit with such alacrity not only because she had the confidence of easily finding another position but also because she was so financially shrewd. She pared costs to the bone through canny shopping and bargaining, through managing the munificence of smitten men, and through her noted ability to elicit "gifts" from her employers.[43] Harrison surely viewed the latter as a well-deserved component of her pay, the customary perks that afforded servants a margin of survival.

Revealing little about her jobs in the United States, she disclosed even less about her modes of transportation, how she moved from one city to another. We can surmise that the bulk of her domestic travel was by train, anticipating her many rail journeys overseas. In the first decades of the twentieth century, the automobile was still a luxury, while US bus service primarily consisted of local routes. Trains, by contrast, were ubiquitous and affordable. Numerous rail companies provided regional services, and national networks traversed the continent.

These trains of course varied tremendously across regions and populations, from the luxurious Pullman cars chartered by upper-class white passengers, replete with dining cars, bedrooms, and valets, to segregated carriers in the South. "Jim Crow cars" actually originated in Massachusetts in the 1830s, when the state's railroad companies began to consign Black passengers to cargo cars. Pryor describes the practice as "a method of racial control that institutionalized segregation as no method of transportation had done before."[44] Nearly a century later, W. E. B. Du Bois cataloged the hardship and indignities that Black travelers endured on them:

> The "Jim-Crow" car is up next the baggage car and engine. It stops out beyond the covering in the rain or sun or dust. Usually there is no step to help you climb on and often the car is a smoker cut in two and you must pass through the white smokers or else they pass through your part, with swagger and noise and stares. Your compartment is a half or a quarter or an eighth of the oldest car in service on the road.... The plush is caked with dirt, the floor is grimy, and the windows dirty.[45]

"WE WANT TO TRAVEL WITHOUT INSULT" is second in Du Bois's list of demands in the open letter to President Warren Harding that ran in a 1921 issue of *The Crisis*, right after "WE WANT THE RIGHT TO VOTE."[46] African Americans had long recognized travel as "a type of currency," Pryor explains, that could be tendered for economic, social, and political gain. To deny them such currency was to exclude them from full citizenship in a nation whose modernity hinged on its transportation revolution. Travel thus became a "critical site of racial and spatial contestation—from leaving home, to arriving away, to using public conveyances in between."[47]

Du Bois elsewhere stated that "the black man is a person who must ride 'Jim Crow' in Georgia."[48] But what about the Black woman? Miriam Thaggert's *Riding Jane Crow: African American Women on the American Railroad* explores the distinct challenges she faced. Thaggert argues that "because mobility is a significant sign of U.S. privilege and citizenship, Black mobility on the nation's railroads became a key issue to access new constitutionally granted rights," and that especially the experience of Black women exhibits "the paradoxical progress and retrenchment, mobility and stasis of Black travel during this period, as Black women determined if and how feminine respectability could travel with them on the train."[49] Many trains offered a ladies' car, which was also a de facto first-class car. However, African American women were routinely denied access to them, no matter their social standing or the ticket in their hand.

Anticipating Rosa Parks's refusal to give up her seat on a bus in Montgomery, Alabama, such travel constraints were early flashpoints for women's protest. In 1863, Charlotte Brown sued San Francisco's Omnibus Line after she was forced off a streetcar, as did Mary Pleasant in the same city in 1866.[50] "Black women," Lynn Hudson states, "beat a path to the California courtrooms to challenge poor treatment on the increasingly segregated streetcars."[51] In Tennessee a generation later, a similar incident in a ladies' car was the precipitating event of *Brown v. Memphis & Chattanooga Railroad Company*. Activist Ida B. Wells likewise sued for redress after being ejected from a Memphis train.[52] As we will see, attaining the privileges reserved for wealthier passengers was a key element of Harrison's travel satisfaction, and she expended considerable narrative energy recounting how she did so.

Yet accommodation was surely her most vexing problem. While by the early twentieth century legally segregated travel was largely confined to the South, segregated accommodation was prevalent nationwide. Many hotels in major cities, including New York, Chicago, Detroit, and Denver, refused to accept Black guests. *The Negro Motorist Green-Book*, published in 1936 as private automobile travel accelerated, was one of a number of guidebooks designed to help African Americans navigate a segregated country. Originally confined to New York but soon offering national coverage, the *Green-Book* compiled lists of hotels along with restaurants, recreational facilities, and other venues that received African American customers. Harrison joined countless other vacationing Americans at popular destinations like Pikes Peak, St. Augustine, Niagara Falls, and the Great Salt Lake, but Black travelers faced many obstacles. In a 1917 profile of Idlewild, the rural Michigan retreat that served the growing African American middle class, Du Bois remarked that "the ever-recurring race discrimination" made for "a puzzling query as to what to do with vacations."[53] African American resorts such as Idlewild, Oak Bluffs on Martha's Vineyard, American Beach in Florida, Lincoln Hills in Colorado, and Murray's Ranch in California gave the affluent some options, but they were too pricey for ordinary people.

Although the organization was far from free of racial prejudice, Harrison found her solution in the nation's YWCAs, which supported women workers moving to urban districts. Along with safe, affordable accommodation, the Young Women's Christian Association looked to offer moral instruction to young women far from home. The first boarding house affiliated with the YWCA opened in 1860 in New York, followed by branches in Philadelphia, Washington, DC, and many other cities. The year 1894 saw the creation of the World YWCA, and by the early twentieth century, there were YWCA branches and affiliates across the United States and worldwide. Facilities varied greatly. Some branches had their own employment agencies. Others offered training in subjects that ranged from domestic skills—an early mainstay—to art and languages. Additional support could take the form of cafeterias, social clubs, childcare, and even medical services. While the organization's mission statement has changed over the years, a promise of "peace, justice, freedom and dignity for all people" has been constant across its iterations.[54]

Their services were vital to Harrison. We know that in the United States she patronized YWCAs in Los Angeles and Honolulu, and there were surely many more, given her reliance on the organization overseas. Her book shows that she visited or stayed at YWCAs in London, Edinburgh, Nice, Rome, Cologne, Brno, Madras, Kolkata, Mumbai, Colombo, Yangon, Kobe, Shanghai, Canton, Jerusalem, and Cairo. The Y furnished her very first bed on arriving in England, and throughout her trip it continued to be her first recourse. "I love here in the Y." (287) she enthused from Kobe, Japan.

Along with providing affordable lodging, YWCAs speedily connected her to networks of mobile women workers. Harrison often applied for jobs through their employment agencies or through other bureaus they vetted. To do so minimized her risk; some agencies were fronts for prostitution, and they could also simply be fraudulent.[55] She took advantage of their educational offerings too: Morris noted, "Wherever it was possible she attended classes at the Y.W.C.A." (ix). They served, moreover, as information hubs. Through them, she could gather advice on finding work, shopping affordably, and moving about safely. In addition to orienting her to current locations they briefed her on ones to come, and she frequently took letters of reference from one Y to the next.

She was far from an indiscriminate user. The local Y often served only as a point of entry before she moved on to cheaper and more agreeable rooms elsewhere. Nevertheless, the organization gave her a steadfast sense of comfort and security. Due to the YWCA, Harrison affirmed from Juan-les-Pins, France, "I feel quite at home everywhere I go" (29). And from Mumbai: "If you have your Y.W.C.A. Directory you always have an adress to go to. Even if you dont stay at the Y. you can come in and get news and there are plenty of girls to talk to" (112). She sounds almost like YWCA publicity copy in reflecting, "It is good to become a Member of the Y.W.C.A. when you start out traveling. I had 3 Y. Cards when I dont know the language my Bible and cards speak for me" (112). In her history of working women in Los Angeles, Eileen Wallis explains that a member's Y card could serve as an employment credential, "an official card of introduction that bore the organization's name to show she was a suitable job candidate."[56] Yet beyond their function as a professional reference, Harrison suggests that the cards "spoke" for her in marking her as respectable and easy to place—just like the Bible she toted. To have her character vouched for was essential to a woman traveling on her own.

"When I came ashore I walked just as straight to the Y. as though I Put up the Building" (148) Harrison relayed at the terminus of her arduous journey to Yangon. We can imagine her doing the same in American cities. However, in sore contrast to her free use of them overseas, her access to YWCAs in her home country was restricted even as endemic housing discrimination heightened her need. Belying the inclusivity touted by its mission statements, many American Ys were reserved for white women. As Thompson wrote in respect to a YWCA in the university town of Grand Forks, North Dakota, "The W. meant white."[57]

At the turn of the twentieth century the Phyllis Wheatley clubs had stepped in to remedy the lack of affordable housing for African American women. Along with shelter, like the YWCA they too offered vocational support and recreational programs with a broad seam of moral uplift stitched in.[58] By 1919 there were forty-nine of them across the United States, and their number continued to grow. Reformers also established the Colored Young Women's Christian Association. The constitution of its pioneering Baltimore branch stated, "There is no class of people so helpless, or who stand so much in need of wise friends as the many young girls who leave their country homes and the care of parents, to seek work in the city.... It is to this class, most especially, that the Colored Young Women's Christian Association wishes to extend it most careful attention."[59] Many others followed. As a counterpart to the YWCA branch in Grand Forks whose "W. meant white," Thompson continued, "the C. could mean colored in Chicago."[60] Some were formal YWCA branches, others were affiliated with the Central Association, and still others were independent. A considerable number grew directly out of the Phyllis Wheatley clubs. Suggesting their eclectic roots, the branch in Washington, DC, was founded by a Black women's literary group, the Book Lovers Club.[61]

Just as African American women managed the Phyllis Wheatley clubs, they assumed control of Black YWCA branches.[62] Glenda Gilmore explains that consequently, "the YWCA provided a forum where middle-class black women could demonstrate genteel leisure activities, teach domestic skills, and act as role models embodying the possibilities of the African American woman."[63] Thus Harrison could have been tutored by other African American women, at establishments where aspiration took middle-class form. Thompson points to their multilayered social structure in describing her experience at the Chicago Y, newly arrived from her home state of North Dakota and intimidated by other residents of a class higher than her own: "Many of the girls were pretty and expensively dressed, no two exactly alike. Some were dark with a black-brown velvetness, two were white-skinned with gray eyes and auburn hair.... They were intelligent, well-mannered girls, with good schooling and from good homes."[64]

When Harrison sailed from Los Angeles to Honolulu in 1925, she gave her home address as Denver.[65] She had spent enough time in that city to save the substantial sum of $800, the equivalent of more than $14,000 today, only to lose it when her bank failed. Throughout the 1920s in the boom years leading into the Depression, the United States averaged 635 bank failures annually, economic tremors anticipating the quake to come.[66] Harrison's falling victim to one such failure was due to more than just bad luck. Since she actively pursued hot labor markets, she regularly entered economies that went through boom and bust cycles.

Denver's racial climate posed additional challenges. The city's African American community was small but flourishing. Men worked for the railroad, while many women were in domestic service, and there was also a well-established Black middle class.[67] However, they succeeded not because of but despite Denver's civic leaders and policies, which were deeply racist. Wielding enormous power, the Ku Klux Klan got a foothold in Colorado in 1920, with membership peaking five years later before abruptly falling off. Denver's mayor and police chief were both members, as were some state judges and congressmen.

The organization's active lifespan in Denver coincided almost exactly with Harrison's residence there. Her choice to spend so much time in Denver at just this juncture shines a spotlight on her travel rationale. While she certainly gravitated to liberal societies, she also chose to live in quite the opposite. Her hunger for new places overrode any reservations their discriminatory practices might have prompted. Perhaps her faith in the efficacy of her coping strategies, coupled with her sheer charisma, gave her the confidence to make these choices. She cultivated a personal geography that, although influenced by racial politics, was not dictated by them.

If she stayed at a Y in Denver, it must have been the Phyllis Wheatley Branch for Colored Women on 2460 Welton Street, since the city's central branch was segregated. YWCAs were especially diverse in the West, a crucial resource for nonwhite populations. In her fine-grained history of the Denver organization, Marcia Tremmel Goldstein states that the patrons of Ys in western cities such as Denver, Los Angeles, San Francisco, and Seattle included "blacks, Indians, Hispanics, Japanese, Chinese, and assorted white populations." "It is likely," she argues, that in western YWCAs the "racial and ethnic dynamics" were distinct and "complex."[68] Their multiethnic character may have contributed to the West becoming Harrison's US region of choice.

In 1921, the Phyllis Wheatley Branch of the Denver YWCA described its facility as comprising a "boarding home, employment department, room registry, girls' work, gymnasium, club."[69] The same year its twelve rooms housed nineteen long-term residents.[70] Soon it also accommodated migrants passing through Denver on their way east to Chicago or west to Los Angeles, as well as, the board reported, "girls ill, stranded or out of work."[71] Its Traveler's Aid Service met new arrivals at the station, and its employment agency placed applicants as maids, cooks, secretaries, and laundry workers.[72] The residents themselves organized many clubs, including the Industrial Girls Club, which created a small lending library of books by contemporary Black writers.[73]

Goldstein notes a telling detail, that the branch also received middle-class tourists. African American travelers had trouble procuring lodging in Colorado, in its capital and mountain towns alike. Black vacationers in the latter were usually put up in private homes. Some thirty miles west of Denver, entrepreneurs in Gilpin County offered an alternative when they parceled one hundred acres of blighted mining land into plots for holiday cabins. By 1926, they had sold over a thousand lots to African Americans from as far away as New York and California, leading to the founding of Lincoln Hills, the only Black resort community west of the Mississippi.[74] In the same town, Obrey Wendell Hamlet opened Winks Lodge, which catered to the entertainers and musicians denied accommodation in Denver hotels after performing in the city's clubs, among them celebrities like Ella Fitzgerald and Duke Ellington.[75] They sometimes performed at Winks, and writers including Hughes and Hurston gave readings, too, inspiring the community's nickname, "Harlem in the Mountains."[76]

Harrison, however, lacked the financial resources—as well as, one would imagine, the inclination—to build a vacation cabin or linger at a resort. In accessing outdoor leisure activities, once more she might have found a solution in the YWCA. The Denver YMCA and YWCA jointly organized a Field Meet and Picnic, from 1922 onward held annually at Rocky Mountain Lake Park and attended by as many as eighteen hundred people.[77] Going further afield, although its programs did not fully develop until after her departure, the Phyllis Wheatley Branch was committed to giving girls and women access to mountain recreation. In 1923 it opened Camp Nizhoni in Lincoln Hills, which over the years offered countless Black girls transformative wilderness experiences. It could have been through the Y that Harrison took the excursion to Pikes Peak she references in describing her days relaxing in the Swiss Alps: "I look back to all the hard hiking I did in the Boundries of Pikes Peak. and are quite contented" (41). She spent enough time in the Rockies to invoke them years later in Turkey. The shepherds made her think of "the Bible times" (61), but the awesome landscape recollected the rugged terrain of the Centennial State. She narrates, "Now we are traveling through cenery like the great Rocky mountains of Colorado we are at a little Station High on the rim of a mountain with the level Plains miles away it is really Grand" (61).

I will close out this chapter with Harrison's visit to the Great Salt Lake, to which she alludes in recounting her impressions of the Dead Sea while on a day trip from Jerusalem. Apparently she found the American lake more arresting, as she complained of the Dead Sea, "It was nothing that made it look dead. not near so heavy and salty as Salt Lake" (74). Harrison never lived in Utah; her closest known cities of residence were Denver and Los Angeles, five hundred and seven hundred miles away. Perhaps she stopped off at Salt Lake City while moving from Denver to Los Angeles—a layover that would put the kibosh on Morris's tale of her limping into Los Angeles, broke and broken, to find succor with the Dickinsons. Or perhaps she saw the lake during an organized tour from Denver that included other major natural attractions, such as the Grand Canyon. Scant weeks into her world voyage, she familiarly commented in respect to her back-to-back tours of Scotland and Ireland, "I always go early on these excursion trips so I get a good seat" (8). The reference to routine practice just as her international travels began proves she had developed the habit of package tours while still in the United States.

At the time, any visit to the Great Salt Lake was brokered through Saltair, the outsized entertainment complex on its south shore. Founded by Mormon civic leaders in 1893, the "Coney Island of the West" was meant to furnish the citizens of Salt Lake with wholesome entertainment. The showpiece was an enormous wooden palace of vaguely Moorish design, standing on stilts far out over the water. In high season, holidaymakers were brought in by the thousands on low-fare trains. Famous for its huge roller coaster and vast wooden dance floor, Saltair also offered a legion of plebeian amusements, the likes of "laughing parlors, a merry-go-round, Ping-Pong parlors, spot-the-stones, a penny wheel, several pool halls, a photography gallery, a penny arcade, a silk hose wheel, and a shooting gallery."[78] Live performances, including vaudeville, were a staple.

Harrison visited the Great Salt Lake at the peak of its popularity, the same time that Wallace Stegner worked at Saltair as a teenager. In "Xanadu by the Salt Flats: Memories of a Pleasure Dome," Stegner evocatively reminisces about his commute home after long shifts of hawking hot dogs: "I always took the last train, the eleven o'clock, perching in my ice cream pants and my straw skimmer where I could watch the city's lights and the dark loom of the mountains rush toward me across the flats. We were buffeted by the night wind and the salt-flat smells.... Whole cars sometimes burst spontaneously into song." Yet while for Stegner, Saltair was a "place of magic," Harrison would have had to navigate this diverse social space far more cautiously than this white boy.[79] As she lay bobbing in the celebrated buoyant waters of the Great Salt Lake, her feet in the air above her head, surrounded by families, couples, and a few solo visitors like herself (although the latter were most likely men), she might have looked forward to returning to a bed and a meal—and a drink—at the YWCA Salt Lake City, one of the few establishments in Salt Lake City to lodge African Americans in the early twentieth century.

In 1947, a *Crisis* editorial echoed the "puzzling query" about vacationing that Du Bois had posed thirty years earlier: "Would a Negro like to pursue a little happiness at a theatre, a beach, pool, hotel, restaurant, on a train, plane, or ship, a golf course, summer or winter resort? Would he like to stop overnight in a tourist camp while he motors about his native land 'Seeing America First'? Well, just let him try!"[80] By then of course, deep in her middle age, Harrison had long since ceased to search for American solutions. Rather than devote any more of her tactics and resources to "Seeing America First," in June 1927 she took her plans global. But before doing so, she passed a highly influential period living and working in Los Angeles, the subject of the next chapter.

CHAPTER 3

Los Angeles Friendship and Patronage

From here I turn to Harrison's relationships with the Foster and Dickinson families in Los Angeles and the support and aid they gave her. Alice Foster and her family, African Americans with Mississippi roots who over the course of three generations worked their way well up the class ladder, befriended Harrison while she lived in Los Angeles in the mid-1920s. Myra and George Dickinson, wealthy white property developers who hailed from Topeka, Kansas, employed her during that period as a housekeeper. Harrison's letters to the Dickinsons comprise the majority of *My Great, Wide, Beautiful World*, while her unpublished letters to Foster give invaluable information about her life and views. Myra Dickinson, George Dickinson, and Alice Foster all facilitated Harrison's travels and fostered her writing. The latter has received far less credit for doing so, scarcely visible in a historical record that foregrounds Harrison's white associates, but other Black women literally archived the texts that prove her contribution.

For a number of reasons, the Dickinsons are disproportionately represented in the following pages. Harrison had a longer relationship, and by all appearances one more meaningful and consequential, with them than with the Fosters. We know a great deal about it, since her letters and book include many fond comments about the pair. Moreover, the Dickinsons' history is far more salient and recoverable, due to their race, wealth, status, and power. They make regular showings, to use P. Gabrielle Foreman's formulation, in "the repositories that preserve the history of those who have the social and economic capital to make their stories resonant."[1] Inevitably to some degree this means I perpetuate the cycle this book is meant to disrupt, in which privileged people, due to prior visibility, remain in focus. However, along with providing greater insight into the specificity of Harrison's experience, looking critically at her relationship with the Dickinsons gives us a more informed view of another, less prominent, Los Angeles family, the Fosters, and helps us evaluate their different forms of patronage.

A leading, persistent element in the tale told about Harrison, in her day and ours, is that her world travels were launched after the Dickinsons "established an income" (x) for her by expertly investing her wages. The source is Mildred Morris's preface to *My Great, Wide, Beautiful World*, which rehearses the events that led Harrison to Los Angeles:

> A large part of her weekly earnings she usually managed to save and she had accumulated eight hundred dollars in a bank at Denver where she happened to be working. Soon she felt she would be able to realize her childhood dream. But the bank failed and Juanita lost her small fortune.
>
> Then a happy turning point was reached. She had sufficient money to buy a railroad ticket to California and went to Los Angeles where she secured a position in the home of Mr. and Mrs. George W. Dickinson.... A real estate broker, Mr. Dickinson invested Juanita's monthly salary in mortgages, until he had established an income of about two hundred dollars a year. With this slender security in reserve, Juanita started out in June 1927 to work her way around the world.

There is much to question about this tale. The place to start is that "slender security." One eyebrow might rise on learning that Dickinson invested her money in his own real estate firm—not entirely an act of benevolence. The other will lift on considering that as a consequence, Harrison suffered a net loss when the stock market crashed in 1929, leading her to hole up for years in France while her finances recovered. But of greatest interest is the segregationist policy with which that investment income is bound up. Reputedly "the oldest firm of subdivision specialists in the city," Dickinson's real estate company pioneered Los Angeles's noxious restrictive covenants.[2] Paradoxically, Harrison's globally scaled mobility was bolstered by a system that curtailed the geographical freedom of countless Black Angelenos, regulations that determined the very neighborhoods in which they could live. And her "true friends" (FL2) personally profited from implementing such control.

This race-based spatial constriction is the dark shadow of Los Angeles's defining expansiveness. At the beginning of the twentieth century, the city's population and territory grew at a breathtaking pace. The number of inhabitants increased by 147 percent between 1900 and 1910, and the rate of physical growth was even faster. Los Angeles swallowed up "previously incorporated communities" along with "thousands of acres of unincorporated land."[3] Speculators bought swathes of desert land to subdivide into housing tracts, peaking in the 1880s and again in the 1920s. In 1923 alone, a new tract opened almost daily.[4] The company that George Dickinson directed with his partner, Frank Strong, was one of the leading developers. Strong & Dickinson specialized in tony neighborhoods to the west and south of downtown. As early as 1925, the firm had 145 "extensive properties" to its name, among them projects such as Ramport Boulevard, Moreno Highlands, Alamitos Bay, and the Silver Lake District. By the time Dickinson died eleven years later, he was said to have opened two hundred subdivisions.[5]

The Dickinsons took an active interest in Harrison's financial affairs, in a manner inseparable from their management of their private household and business enterprise alike. In the course of describing their history together, George itemizes her portfolio—egregiously—in a letter to *Atlantic Monthly* editor Ellery Sedgwick, which Sedgwick chose to publish:

> Juanita worked for us for four years prior to 1928, which were prosperous years prior to the crash of 1929. Domestic wages were very high and we paid her $75 a month, including room and board; and, as you know, Juanita never spent a cent unless she absolutely had to and saved every cent that she could get hold of. She always made friends with whomever she worked for and a great many gifts were given to her in the way of dresses and clothes. From time to time we purchased for her a $900 trust deed (mortgage) that bore 8 percent interest, a $450 trust deed at 8 per cent, a $650 trust deed at 8 per cent, and a $550 trust deed at 8 per cent. That made $51 interest quarterly, or $204 a year. After the crash in 1929 the trouble commenced; we reduced the interest rates. Two of the trust deeds fell down entirely. Now, instead of having her money invested in what we supposed were good mortgages, drawing good interest, she has but little interest, with vacant lots that at the present time are an expense to her.[6]

As his statement, "we reduced the interest rates," reveals, the investments made for Harrison were in their own company: the "very high" wages he and Myra paid her were funneled right back into the family business. This crucial fact is always overlooked in discussions of the assistance they gave her.

Common in California, the "trust deeds" to which Dickinson refers attest that although the landowner paid the mortgage on a property, the holder of the deed had an interest in it, potentially generating passive income. Complementing those four deeds (now two), the "vacant lots" she was losing money on at the time of the letter prove that in 1935 Harrison was the actual owner of more than one plot of land. Naturally enough, given that they owned the deeds in which she invested, the Dickinsons served as the administrators of her portfolio. During her travels, they forwarded her the interest she had due and occasionally advanced her funds against future gains; they also handled her property taxes. But for all their support, apparently they were not inclined to cut her a break during the hard times that followed on "the crash in 1929," to give her an interest rate higher than the market dictated.

Los Angeles, as Strong & Dickinson's proliferation of tracts manifests, was by design a horizontal city. Lawrence Culver explains that developers did not look to create a single urban center. Instead, in a "radical departure from traditional patterns of urbanism," they planned a multitude of low-density suburbs, connected to jobs and services by the pioneering Pacific Electric train and streetcar network.[7] Promoting health and recreation, such diffusion was conceived as a salubrious alternative to the crowded tenements of northeastern and Midwestern cities. In 1911, Los Angeles had less than half the population of Boston, even though it was four times its size.[8] The majority of the inhabitants of the nation's "first truly suburban metropolis" lived in houses rather than apartments, many of them houses they owned. Among large cities, Los Angeles's rate of homeownership was one of the highest in the country, a powerful draw for migrants.[9]

Echoing Morris's formulation of her Los Angeles move as a "happy turning point," contemporary critic Owen Walsh maintains that the advantages Harrison enjoyed in California, including her good job, helpful employers, and—especially—ability to buy property, were "easier for a woman in her social position to find in Los Angeles than almost anywhere else in America."[10] Property investment was indeed the path forward for the city's most successful African American entrepreneurs. Ordinary Black Angelenos were not excluded either: More than one in three owned their home.[11] W. E. B. Du Bois enthused that in Los Angeles, "the beautiful homes lay low crowding on the earth as though they love its scents and flowers. Nowhere in the United States is the Negro so well and beautifully housed."[12]

Yet this was only opportunity of a relative nature, compared to the rural South and the urban North.[13] Moreover, as "racial animus accumulated and patterns of racial oppression consolidated," to quote Scott Kurashige, this opportunity began to narrow.[14] Its high rate of Black homeownership notwithstanding, Los Angeles implemented restrictive covenants. As Gene Slater formulates the development, "Newly established all-white real estate boards, including the Los Angeles Realty Board, the largest in the country, organized the industry and came to control the vast majority of home sales ... to control *whose* money could buy a home."[15] By the 1920s restrictive covenants were ubiquitous in Los Angeles.

The majority of the new tracts prohibited sales to nonwhite people—or rather, while they could buy lots and buildings, they were not allowed to live on the property. For many Americans the California dream meant living in an exclusively white neighborhood, and the subdivisions, segregated from inception, enabled them to realize this dream. White Angelenos could maintain a veneer of tolerance without having to live among racial others, save those employed as domestic help; crucially, servants like Harrison were exempt from the restrictions. Kurashige characterizes the development as a "closing of a frontier" for African Americans and Japanese Americans.[16]

Strong & Dickinson was a prime agent of this closing, as one of the first firms to deploy restrictive covenants. Reporting on the minutes of the Los Angeles Realty Board's July 1911 meeting, Laura Redford discloses,

> The Title Insurance & Trust Company approached LARB members Strong & Dickinson, who had written "the premises shall not be conveyed or transferred to any person or persons other than of the white or Caucasian race" into property deeds. The title company was concerned that the legality of a racial covenant was not yet settled by the Supreme Court and wanted "the Realty Board [to] take the matter up and carry it through to a finish and have the question settled once and for all."[17]

In other words, the company implemented such covenants well before the 1919 ruling in Los Angeles Investment Co. v. Gary made them legal.[18]

Indisputably, Harrison's annual investment income—those $204—was the economic pillar of her first two years of travel, a regular source of income from the United States that while not enough to allow her to forego wage labor altogether, subsidized her activities. Consider her pleased report from Seville: "Living is very cheap and the dollar high which pleases me" (200). That she should travel on income from land speculation does seem apt. Her world venture, predicated on advantages gained by cycling through different physical locations, was boosted by trading actual plots of land. Yet even as she wrote appreciatively about Nice, France, as a city where she could "always get a comfortable and Home like place to stay for here you never think of your color" (21), her international travels were underwritten by funds derived from discriminatory housing practices. The Dickinsons were committed to supporting their former housekeeper in her private quest for freedom, but they were also lead architects of what one Black Angeleno figured as "invisible walls of steel."[19]

This seeming contradiction is a symptom of the complex nature of Harrison's relationship with Myra and George and their surprising role in her journey and authorship. I certainly have reservations about accessing Harrison's life through people who were not just immersed in, but actively helped establish, Los Angeles's racist institutions. The Dickinsons are near emblems of white privilege. One of the last photographs of George shows him beaming in plus-fours on the golf course of the notoriously discriminatory Los Angeles Country Club.

Nonetheless, to give the Dickinsons less than complete attention because of the unequal power dynamics is to do Harrison's emotional reality a disservice. These are the people whom she repeatedly identified as the most important in her life. It is hard to overstate how effusively she wrote about them. "From my experience not another human in this world are as Kind as They are" (FL2), she stated. To get a letter from them was "like getting home." What greater endorsement, from a woman who once bitterly remarked that she had never "had a home to return to" (221)? Understanding the Dickinsons helps us understand Harrison. At the same time, I see this work as a corrective, recasting with greater precision the widely circulated story of their support.

Harrison was one of thousands of Black Southern transplants in Los Angeles, the majority from Texas, Louisiana, Mississippi, Arkansas, and Oklahoma. Since many hailed from urban areas, they were more prosperous than the average migrant.[20] Los Angeles actually had fewer job opportunities for African Americans than did industrial cities in the Midwest and the Northeast. They were drawn to Los Angeles not by the promise of factory jobs but rather by the "generally good conditions" of southern California.[21] Along with the prospect of being "well and beautifully housed," these attractions included an agreeable climate, stunning geography, and abundant recreation. California was far from the free haven that abolitionists had once touted, but, Lynn Hudson maintains, confidence that it could become such a haven "drew migrants hoping to turn the belief into a reality."[22]

Knowing Harrison, LA's cosmopolitanism was more of a draw than its African American community or the prospect of homeownership. Mark Wild explains that "fast-growing metropolises of the West," chief among them Los

FIGURE 1. George W. Dickinson at the Los Angeles Country Club. Source: "George W. Dickinson golfing at the Los Angeles Country Club, *Los Angeles Times*," 1935. Los Angeles Times Photographic Collection. UCLA Library Digital Collections.

Angeles, offered a "new modern industrial American community" that featured "cosmopolitan areas where residents of all races mingled in the streets and public establishments."[23] Los Angeles, moreover, was one of the earliest "World Cities," proximate to both the Mexican border and the Pacific Ocean.[24] Living there prepared Harrison for the dynamic societies that would soon captivate her overseas.

Adding to the attraction, southern California already had a reputation as "the playground of the world"—exactly what she liked.[25] It was aggressively boosted as "an exotic cultural and ecological wonderland."[26] Culver argues that while Los Angeles is often linked with the ascent of consumer culture, what made it truly distinct was its beckoning of "leisure as a permanent way of life"—that is, a new "lifestyle."[27] He places the city at the "frontier of leisure," and as Alison Rose Jefferson demonstrates in *Living the California Dream: African American Leisure Sites During the Jim Crow Era*, African Americans were very active on this frontier.

For all the enticement of relaxed living, the population of Los Angeles would never have soared as it did were it not for the city's capitalist opportunities, the myriad ways it promised money. Ambitious women and men of all stripes flocked there to profit from ventures in agriculture, industry, business, and real estate. Their numbers included many who, like Harrison, went in search of well-paid service jobs. In 1913 the head of the local YWCA's "Vocational Department" rued, "This office is daily embarrassed beyond words by hordes of women who have come here from the east without money or friends, expecting to find fine positions the moment they stepped off the train."[28] To accommodate the surge, the organization founded a boarding house in 1905, and the Mary Andrews Clark Memorial Home, also affiliated with the YWCA, began receiving residents eight years later.

These establishments, however, were for white women only. In Los Angeles, Harrison faced her perennial American problem of finding accommodation in a segregated town. Eloise Bibb Thompson, a local journalist, wrote about the painful difficulty that single Black women had procuring safe housing there.[29] On first arrival, Harrison may have turned to the African American branch of the YWCA. Or perhaps she sought refuge at the Sojourner Truth Home, opened in 1914 by the Sojourner Truth Industrial Club. The boarding house was hailed as "filling a long felt need in Los Angeles. Giving a protecting shelter to self-supporting women and girls in the city and to strangers." It also offered instruction in subjects such as "Domestic Science," "Bible Studies," and "Physical Culture"—all of which would have interested Harrison.[30]

Just when she alighted in the city is unclear. Morris's succinct tale in the book's preface of her flight from Denver to Los Angeles omits a major event that disrupts its timeline along with its redemption plot: in January 1925, she had sailed from Los Angeles to Honolulu, where she stayed for half a year. Suggesting how recently she had left Colorado, the passenger manifest of her outbound ship lists her hometown as Denver.

That the Hawaii sojourn is absent from the preface can be variously accounted for. Perhaps Harrison neglected to tell Morris about it. Perhaps it was edited out for brevity. The likely cause, however, is the inclination of both editor and publisher to highlight the Dickinsons' savior role, a scenario that the majority of Harrison's reviewers were primed to endorse. Her visit to Hawaii put this in question. Having the resources to go there so soon after the financial hit she took in Denver shows she was less hapless than Morris implies.

In respect to "true friends," Harrison once confided to Foster, "No doubt you found them early in your life. But I not until 1924" (FL2). Underscoring the Dickinsons' impact, this statement also gives us a starting point for their association, 1924, a year that George's *Atlantic Monthly* letter corroborates. Thus it is likely that she worked for them between leaving Denver and her January 1925 embarkation in Los Angeles for Honolulu, although perhaps not for a long enough period to conceive Los Angeles as her home. Her Pacific voyage might even have been inspired by hearing about the Hawaii vacations the couple had previously taken—no less than three of them—that made them some of the earliest Americans to enjoy the territory's new resort facilities.

On Harrison's return to the continental United States, after landing in San Francisco she headed south to enter or reenter the Dickinsons' household. She sailed for England two years later, but not before briefly taking another job in New York. While not literally untrue, Dickinson's statement, "Juanita worked for us for four years prior to 1928," is misleading. She was in Hawaii in 1925; her world trip began in June 1927; and she worked in New York between leaving Los Angeles and embarking for Europe.

George and Myra's achievement is certainly not as extraordinary as Harrison's. Nevertheless, their lives do exhibit a formidable "American Dream" trajectory in their journey to membership in Los Angeles's ruling class. Both their families had roots in the northern town of Malone, New York (incidentally, the childhood home of Almanzo Wilder and setting of *Farmer Boy*), and both are characterized by inveterate westering.

In 1849, Myra's maternal grandfather, an "adventurous spirit," abandoned his family to scarper to California.[31] Her parents, Charles and Martha Kellam, left New England to settle in Topeka just prior to Kansas statehood, where Myra was born; she was "known from childhood to the old settlers."[32] Charles worked as a pharmacist and went on to own a drug store. The Kellam family had a live-in servant, twenty-two-year-old Maggie Herr from Germany.[33] Myra grew up living with domestic help, although, due to shared racial identities, on more equitable footing than she would later be with her African American housekeepers in Los Angeles.

When she and George married in 1882, the young man had already clocked three years as a "cashier for the Santa Fe"—that is, the Atchison, Topeka and Santa Fe Railway, which linked Kansas with New Mexico.[34] At the time his father, Colonel William Green Dickinson—once a politician in Malone and then a civic leader in Topeka—was director of the Santa Fe's real estate branch, the Land and Town Company. In that capacity he founded a prodigious eighty towns along the Santa Fe line. The Santa Fe had broken the Southern Pacific Railroad's monopoly by completing a track to Los Angeles over the Cajon pass, sparking a fare war and the region's first real estate boom. The scale of development was outsized, and the Dickinsons cashed in.

To do so, Colonel Dickinson and his adult children relocated to the nascent community of National City just outside San Diego, the site of the company's new headquarters. Originally Spanish grazing ground, the settlement saw its name change from El Rancho del Rey to Rancho de la Nación when Mexico won its independence and then change again to National City after the border shifted south. The elder Dickinson spearheaded three major projects there: damning the Sweetwater River, establishing San Diego's first commuter rail line, and platting the town of Chula Vista. He envisioned the latter as a pastoral community of citrus orchards and fine homes.[35]

Both families collectively, it appears, decided to settle in National City, with two generations of Kellams and Dickinsons making a series of moves there in the late 1880s. Myra's parents moved to National City in 1887, the year it was incorporated; her father and sister would later manage the bustling International Hotel, which put up the managers and workers who developed the town.[36] That same year Colonel Dickinson had an opulent eighteen-room Queen Anne Victorian mansion built for himself. George's sister Mary Boal moved into the house when the Colonel died in 1891, and her husband took over his position at the Land and Town Company. George's older brother, Wallace, built his own palatial home across the street. George came from California developer royalty.

By 1893 he had established his own property firm, George W. Dickinson & Co., perhaps with the help of a bequest from his father. Some seven years on, it was characterized as "the well known and reputable real estate brokers of San Diego."³⁷ (His future partner in Los Angeles, Frank Strong, also started out in San Diego.) Myra and George's three children were born in National City, including Charles, who died as an infant; Martha, who at twenty-one succumbed to a long illness; and William, who later worked with his father and about whom Harrison fondly wrote. Within a scant few years, the booming real estate market that drew the Kellams and Dickinsons to National City spectacularly crashed, leaving at least one of the town's founders destitute. Nonetheless, many of the family lived out their days in National City and are buried there.

That they were so locally entrenched may have contributed to Harrison settling in the greater San Diego area in the early 1950s. I hope that my laying out their history will help future researchers discover more about her life in southern California, whether her 1920s tenure with the Dickinsons or the obscure period after she repatriated. At the same time, the stark contrast between Harrison's and the Dickinsons' California histories is itself instructive. Not only through westering but also through family connections and legacies, the somewhat ordinary Dickinson clan could move from average circumstances to exceptional ones, to great wealth and influence in California. Far more mobile, creative, forceful, and determined, Harrison had to make her way without this foundation of generational family support. Instead she was intermittently boosted by forms of patronage and aid.

At the close of the nineteenth century, Myra and George's course took an international detour—and this is where their experience starts to resonate more profoundly with Harrison's. Having lived scarcely ten miles from the Mexican border in National City, as of 1900 George was posted at the US consulate in Acapulco. The reign of President Porfirio Díaz was at its height, making for a very enterprising time for Americans in Mexico. On resuming the presidency in 1884, Díaz had modernized Mexico's financial and industrial sectors and courted foreign investment. His controversial policies supplied US businesses with a labor force and Mexico with greater infrastructure. American investors flocked south to take advantage of the new opportunities, advised by officials like Dickinson. Some of the ventures were quixotic. Dickinson was tasked, for example, with investigating the prospect of making Mexico a "market for purebred cattle" imported from the United States; American breeders promoted their self-interest in maintaining that Mexican cattle were as much in need of racial uplift as Mexican people.³⁸ His conclusion was disappointing: "I have to report that, after consulting with such ranchmen as come within my reach, there is no demand."³⁹

The consulate gig does not seem to have been full-time, as during this period he and Myra also set up housekeeping in Los Angeles, where he joined forces with Strong. Myra, too, as a profusion of notices in the *Los Angeles Times* attests, was busy buying and selling plots of land. For about five years the pair went back and forth between Acapulco and Los Angeles before settling down in the latter. The year 1909 sees them at 423 Andrews Boulevard—soon renamed Lafayette Park Place—where Myra was to live for the next fifty years. When they first bought the house, they still were not fully established: the following year, a forty-four-year-old white nurse from Pennsylvania was living with them as a boarder. The arrangement indicates that even in their late middle age, they needed extra income to yield a margin of comfort.

This was the property where Harrison "lived in" with the Dickinson family in the mid-1920s, working as housekeeper and cook. The Lafayette Park neighborhood was fast becoming a fashionable residential area. By 1929 its transition was complete, as the Town House Hotel, Felipe de Neve Branch Library, and, most importantly, Bullock's department store were all erected that year. So elegant and visually striking as to feel as much museum as store, Bullock's sported an entrance that, heralding the future, was approached not on foot but by car. Boosters gushed over investors having the confidence in the city's growth to build so far from downtown. His fortunes surging with the city's, George Dickinson came into his own. In 1920, he made William his partner in what was now a family firm, Geo. W. Dickinson & Son. A notice promoting the enterprise assured clients of "the same conservative, substantial, and yet aggressive policies which have always been associated with Mr. Dickinson's identity."[40]

By then the family had begun their long-standing practice of employing a live-in African American housekeeper from the South. While in 1910 their household included that paying boarder from Pennsylvania, in her stead a decade later was Labada Robinson from Arkansas, working for wages. Harrison was next, succeeded in 1930 by Maude McCalph from Missouri.[41] Having a full-time, Southern Black housekeeper was a sign of status in California, announcing that the Dickinsons had made it into the upper class. The presence of women like Robinson, McCalph, or Harrison in their household made a very different social statement than had that of the German Maggie Herr in the Kellams' Topeka home. That said, the Dickinsons' relationship with Harrison, an attachment that endured long after she left their home, city, and country, was surely unlike those they had with their other housekeepers. In her book, Harrison fondly describes retrieving a family portrait they had sent her in Burma: "I went to

Cooks & Son. and got the Happiest shock when I open your letter and saw the Photo.... The Photo. looks swell on my walnut dresser. and each One in it have their own beauty" (149). Their friendship was real.

The Dickinsons' own love of travel must have nourished their support of her quest. In turn, Harrison may well have been inspired by their activities. Even as accounts of the couple overemphasize their influence as her financial advisors, they are entirely neglected as possible travel role models. During their first year of marriage, 1893, the Dickinsons traveled to Chicago from National City in order to "do" the Chicago's World's Fair, as a Topeka newspaper reported.[42] They went on to become major travelers and even adventurers, a predilection seeded by their term in Mexico and the regular voyages they made between Acapulco and California.

They especially enjoyed long tours by car, pioneering that private mode of travel. In 1913, according to the *Los Angeles Times*, they took a "two months' motor trip through the east," and some eleven years later, they made an even more ambitious road trip that crossed into Canada.[43] One Ruth Higgons hosted a jolly "Farewell Party" prior to taking a "seven weeks' trip east" with Myra and George and their niece Mary Ann (who as we shall see would later promote Harrison's book). The plan was to go by train to Syracuse, New York, and buy a car in that city. They would then tour both sides of the border, stopping off in Atlantic City, Philadelphia, New York, Montreal, Quebec, and Niagara Falls. From there they would drive to Chicago and freight the car by rail back to California.[44]

Of course such excursions were comfortably padded. The Dickinsons' budget was capacious, and they had a home, business, and bank account to return to. That seven-week driving tour, for example, was clearly an upper-class indulgence, requiring cross-country train tickets, the purchase of a car plus freight costs, and weeks in hotels. Nonetheless, it was also rather hard-core. Driving their car, drawing up their itineraries, and making their bookings were all physically arduous or logistically complex. Traveling in this manner makes the Dickinsons quite early and intrepid auto tourists, and there are surely other ventures lost to record.

Their most evident link to Harrison's enterprise is the "World Jaunt" on which they embarked in December 1925. George and Myra sailed "under the direction of the D.F. Robertson Travel Bureau" as members of "one of the largest parties to leave Los Angeles on a tour around the world."[45] Reporting on the itinerary of the "globe-girdling trip," the *Los Angeles Times* imparted, "They will visit Honolulu, Japan, China, the Philippines, Malay Peninsula, Burma, Ceylon and Egypt. On arrival at Marseilles, they will proceed by automobile through Europe." The party would "motor extensively" on the continent before "crossing over to the British Isles."[46]

The Dickinsons took this trip during the course of Harrison's employment at their home, and she began a "World Jaunt" of her own scarcely a year after their return. Their ventures share considerable common ground. Commencing with a transatlantic rather than transpacific voyage, Harrison visited, in the reverse order, all the main destinations that they had—save that she was in India rather than Malaysia—indicating how by the 1920s, around-the-world journeys had assumed a rather fixed profile. It also demonstrates that despite the class gulf between them, she and the Dickinsons could enjoy the same places. Modeling travel of this scale, their around-the-world voyage might have encouraged Harrison to plot circumnavigation herself, or to conceive an already planned venture as such. But there the likeness ends. While their journey was executed within several months under the auspices of a high-end travel agency, she traveled independently for years, interpolated with paid labor and a stretch of true expatriation in France.

In F. Scott Fitzgerald's 1925 novel, *The Great Gatsby*, Nick, Gatsby, Jordan, Daisy, and Tom get a hotel suite in uptown Manhattan overlooking Central Park—just for the day—to cool off during a heat wave. "The notion originated with Daisy's suggestion that we hire five bathrooms and take cold baths," Nick recounts. "Each of us said over and over that it was a 'crazy idea.'"[47] That they should whimsically "hire" hotel rooms of an afternoon is posited as a mark of true extravagance, the lifestyle of the very rich. Yet this is exactly what Harrison did in Los Angeles, in real life, on a maid's salary. She informed Foster, "One of the plannes in my life was when ever I really desired it to remain in bed a ½ day or all day and have my meals served to me by a nice maid or Waiter and in a nice way" (FL2). Therefore, she explained, "When [in] Los Angeles when ever I took that desire I always went to the Hotel Baltimore and took a nice two dollars a day room and there was served by a pleasant Waiter." Despite her thrift, she knew how to spend money on herself when she needed to, to take a break from service to be served.

Her patronizing this particular hotel is still another example of her habitual occupation of storied places. Baltimore Hotel, designed by the same architect of what became Hugh Hefner's Playboy Mansion, was in the heart of the city's financial district, about two miles from the Dickinson home. Over the years, it was the site of a number of headlining murders and suicides. It also served as an operation base for the deadly *Los Angeles Times* bombing on October 1, 1910, with the perpetrator assembling in one of its rooms the explosives he planted hours later. The building still stands amid other art deco edifices, having been converted into low-income housing after a period of disrepair.[48] The "Baltimore Hotel" marquee can be glimpsed in a 1970s promotion video for the Rolling Stones' *Exile on Main Street* project.

In its day, the Baltimore Hotel was advertised as "up-to-date" and "absolutely fireproof," with its "cafeteria dining room" a further draw.[49] Cafeterias were a new trend in the United States, and while the word itself—Spanish for "coffee shop"—was imported from Mexico, Los Angeles has been claimed as their birthplace. Along with the promise of "Food That Can Be Seen," their selling points were convenience, speed, and cleanliness.[50] We can envision Harrison enjoying this modern form of dining with the likes of a breakfast of "Luscious California Sliced Peaches, Crisp Brown Waffles with 2 pieces of butter and pitcher of syrup, and Cup of Coffee with pure cream," purchased for thirty cents.[51]

Her ability to indulge in the day room practice confirms George Dickinson's contention that he and Myra paid her well. However, it also suggests that for all their good relations, she could find "living in" with them wearing. Occasionally renting a room for the day established a middle ground between a private domicile and an around-the-clock workplace. This requisition of personal freedom and space anticipated her eventual decision to leave the Dickinson household altogether.

The more significant expenditure she made in September 1925, for a plot of land in Santa Monica, further demonstrates her willingness to spend when opportune—in this instance, with the aim of future gain rather than short-term gratification. The purchase was made relatively early in her LA residency. Others must have followed, since by the time she left California she owned a set of lots. Bought from an R. S. Shepherd and Ruth Shepherd for ten dollars, the property was on Olympic Boulevard, a straight 1.6-mile shot to the Santa Monica pier.[52] It is typical of Harrison's tastes that the land should be in a developing resort town by the sea. Maybe that lot was just an investment, but maybe she envisioned it as a place to settle down after tiring of travel—to use her own expression, a "little future nest" (311). That she had property there may have contributed to her need to clarify, from Honolulu in 1936, that she did not intend to return to Los Angeles.

There is also the question of whether the lot had a restrictive covenant—legal for Africans Americans to buy, illegal for them to live on. We do not know how long Harrison kept the Olympic Boulevard property, but still in the late 1930s, most of the plots vacuumed up by speculators had virtually no value. It took many years for the Los Angeles property market to recover. Eventually though, she did sell her holdings, as at her death she owned no real estate. An office building erected in 1972 stands on the Olympic Boulevard site today.

Even after her world journey began, in many ways Los Angeles continued to serve as Harrison's home. Investment in city properties subsidized the first stage of her travels, friends in Los Angeles aided her during them, and her ties to southern California supported her in more subjective ways, too. Overseas she represented herself as a Californian, and she may well have felt herself to be one. Periodically offering her harbor, southern California was one of her longest places of residence, and making a personal claim on the state was both a tactical maneuver and an expression of real affinity.

Passing comments in *My Great, Wide, Beautiful World* prove a sense of kinship, such as Harrison noting that "many from Calif" (19) were interred at the American Cemetery in Paris or, on discovering a fellow traveler hailed from Fresno, anticipating it to be a "joy to me" (11). Yet the identity was also strategic, as she freely admits. Culver notes that Los Angeles's film industry gave the city, and by extension the state, "national and global visibility that was unsurpassed."[53] Attesting to such visibility, Harrison remarks from Spain that "all the Picture House show the Calif. Films I am glad I choosed Calif. for my home before I left as every one know it" (192). California lent her recognition and prestige. More specifically, her book shows, it facilitated sociability with other modern women. She recalls of two English girls, "When I told them I was from Los Angeles they thought it just wonderful" (9); reports from Alicante, "Elche have many flappers and all are Hollywood fans" (192); and notes in Zurich, "The Girls think it so wonderful I am from far away California and all are so Lovely To me" (46).

"Far away California," at once familiar and glamorous, boosted Harrison's image overseas. Its self-inventive modernity made it a fit home for her, mirroring her own practice. Her book, moreover, has a salient California dimension in that it is largely composed of letters she sent to people living there. While not written in or about California, it was addressed to Californians, contributing to why I have in the past assessed it as regional memoir. At some level, when Harrison was narrating her travels, California was on her mind.

What else do we know about her life in Los Angeles? A goodly portion of her free time was spent at church. She relished the Hollywood Passion Theatre, the outdoor religious spectacle that she preferred "a thousand times" to the Oberammergau passion play, which in her view did not "come up to Hollywood at all" (260). For more regular fare, she relied on Aimee Semple McPherson's Angelus Temple in Echo Park. Dedicated in 1923 and still standing, the megachurch could be reached by foot from the Dickinson household. McPherson was one of the first modern media celebrities, serving up exuberant theatrics underpinned by conservative beliefs, every weeknight and several times on Sunday. Some of McPherson's racial views and policies were pernicious—she established a segregated church for Black worshippers—but Harrison reveled in her enterprise nonetheless.[54] Her enthusiasm for it helped sustain her California identity overseas. She apprised Foster, "Well I have heard Preachers in the north and Preachers in the South Preachers in the East and Preachers in the West But Sister McPherson is the Preacher that Preaches the best. My temple friend send me the Foure-square maz gine" (FL2). Who was that "temple friend," I wonder, mailing off church magazines to France?

If Harrison kept to her usual habits she also spent ample time outdoors, enjoying the city's parks and beaches—perhaps defiantly so. Even as bountiful leisure was central to the allure of southern California, many locations and activities were restricted to white residents. In 1925, for example, Los Angeles abruptly segregated its plethora of public pools. Jefferson argues that in California, leisure "became a site to separate, segregate, control, and regulate people, places, and opportunities." Consequently, leisure also became a focal point of civil rights activism for Black Californians.[55] On a lifelong quest to optimize pleasure and novelty, a lover of the outdoors and all kinds of recreation, Harrison must have been an energetic participant. "Its my life to see and enjoy" (FL2), she insisted, and her book demonstrates that she routinely claimed spaces and experiences that others would forbid her.

Yet despite all that it had to offer—or cede—she was not fond of the city, alerting Foster, "If I ever come to Los Angeles it will be only to visit the Temple. And to see my People but not to stay as I did not like the Los Angeles People" (FL2). Her aversion stemmed, ironically enough, from its uprooted populace. She clarifies, "I dont mean the ones that were born in Los A. because you meet only a few I mean the ways that they get after reaching there" (FL2). In her view, moving to southern California changed people for the worse. For all the delight she took in multicultural societies and for all her own hypermobility, Harrison often voiced a preference for communities with abiding local populations, something Los Angeles had in short supply.

The Dickinsons and the Fosters, however, must have counted as local in her eyes. The older generation of both families had migrated to California at the close of the nineteenth century, and the second and third generations—with whom she was on familiar terms—settled there permanently. Among them she could count many "born in Los A."

While Juanita certainly has the more arresting migratory history, her friend Alice also traveled a considerable distance from her first home. She was born Alice de Vough in Mississippi scant years after the Emancipation Proclamation abolished slavery. Her father was from South Carolina and her mother, from France. Her somewhat older husband, Lawrence R. Foster, was also born in Mississippi. The couple left Mississippi some time prior to December 1898, the month their first daughter, Alice, was born in Texas. They were living in Houston in 1900, with Lawrence employed as a barber, but they had settled in California by the birth of their second daughter, Edna, the following year. This history makes them one of the earliest African American families in Los Angeles. At most, Harrison was ten when Alice Foster left Mississippi, so their friendship developed outside the South. Perhaps Foster's presence helped draw her to Los Angeles, or perhaps their shared origins led to acquaintance upon arrival. Through their relationship Harrison maintained ties with her native state even while exploring new terrain.

By her mid-sixties Foster was widowed, living in Pasadena with her daughter Alice Cunningham and her young family. Famous for its wealth, the City of Roses was also notorious for its racism. Hudson quotes a new migrant's rebuke: "The only difference between Pasadena and Mississippi is the way they're spelled."[56] It took decades for activists to get its segregated swimming pool policy overturned—only for the city to then close its flagship pool.[57] Pasadena's endemic discrimination might account for why, after only a couple of years there, the Fosters moved back to central Los Angeles, settling in the African American district along the South Central Avenue corridor.

The family were members of LA's "Black bourgeoisie," women and men who "in spite of their working-class wages" had "middle-class aspirations," to quote Jefferson.[58] For African Americans in the early twentieth century, putatively unskilled labor jobs did not necessarily signal working-class status.[59] The censuses do not list an occupation for Foster, but in 1940 her younger daughter worked as a maid and her son-in-law (likewise born in California to Mississippi parents) was a city school janitor, having previously been a driver and auto mechanic. Heralding a middle-class shift, Alice Foster Cunningham was a typist.[60] They were on familiar terms, moreover, with Miriam Matthews, California's first professional African American librarian. Reflecting their attention to racial advancement and professional success, family papers include a 1920s magazine advertisement for Parkridge Country Club, promoted as "the largest country club in the world owned and controlled by black Americans."[61] The family's preservation of the advertisement indicates how important they viewed the achievement.

Harrison began writing Foster shortly after the 1927 launch of her world voyage and continued after its 1935 close in Honolulu; Foster died several years later. Between 1927 and 1929 alone, she sent her at least eleven letters. Their correspondence does not look to have been especially ardent. Harrison addressed her older friend as Mrs. Foster and signed her own name in full, and she wrote less frequently than Foster did (FL1). Rather tactlessly, having informed her that to save on postage she only wrote regularly to the Dickinsons (FL1), two months later she commented that nothing other than the prospect of seeing the Dickinsons would tempt her to pass through Los Angeles again (FL2). Yet the letters she sent Foster were hardly short updates—a dispatch from Nice ran to three thousand words.

In respect to their caretaking of the letters she sent them, the Dickinsons' aim was to support Harrison personally. "My People at Lafayette had saved every one that I might have them to keep for a diary" (FL), she explained. They kept the letters so that later in life, she could reminisce about her travels. Foster and her family, in contrast—perhaps aware of the social significance of her achievement, akin to that Black-owned country club—kept them for posterity. Shortly after the publication of *My Great, Wide, Beautiful World,* Foster loaned the letters still in her possession to Matthews, and some seventy-five years later, her granddaughter, Ann Cunningham Smith, donated them to the University of California, Los Angeles.

The Ann Cunningham Smith collection includes just three letters in Harrison's hand. Dense epistles written in ink on lined yellow paper, two of them were sent from Nice in 1931 when her book was in the planning stage, and another from Honolulu in 1936 shortly after it was published. Their paucity is a direct result of Foster's alacrity. Referring to Morris and the book project, Harrison had asked Foster to send back to Paris all of the letters she had from her, "as the writer may find something in them to help a little."[62] Foster swiftly complied. "Thanking you again for the well kept letters," Harrison wrote her some time later, "which you know not another woman could done that she would have missed placed them and not found them. also thanking you for sending them so promptly" (FL2). Unfortunately, Foster's service hollowed out the archive, since despite Harrison's promise that she would get the letters back—an assurance she made in 1931 and then again in 1936—she never did. (Indicating the value she placed on them, Foster appears to have inquired after their fate a full five years after relinquishing them.) The Smith collection includes eleven carefully filed empty envelopes, the contents of which were presumably sent off to Paris. A tantalizing remnant, the set of envelopes—from Nice, Paris, Rome, Naples, Brno, Colombia, Darjeeling, and London—indexes the lost epistolary exchange of these two Californians from Mississippi.

However, the letters we do have are invaluable. While portions of them resemble the travel narrative in *My Great, Wide, Beautiful World*, unlike the book they also comment on key subjects that include Harrison's family, childhood, racial community, religious experience, and relationship with the Dickinsons. They describe, moreover, the process whereby Harrison and Morris compiled the manuscript in Paris; offer details about the book's reception in Los Angeles and Honolulu; and voice Harrison's ambivalence about authorship and fame. Whereas the letters that form the book, reporting on her travels, are firmly located in the present, the letters to Foster include recollections of her youth and early itinerancy in America. Harrison was especially confiding about her vulnerable girlhood in Mississippi. Their mutual Southern upbringing prompted memories and observations she was less likely to share with the Dickinsons.

The Foster letters serve as a useful control for evaluating *My Great, Wide, Beautiful World*, helping us see the constructed nature of the author's narrative persona. Unlike in the book, Harrison explicitly speaks as an African American, remarking, "Its just in our race we will be kind up until we think that one count on us as the only friend then we let you fall" (FL2). Moreover, as the same comment shows, a far less insouciant personality is on display. While there are still expressions of joie de vivre, she can also sound bitter, anxious, or defensive, another indication of the relative ease she felt writing someone she knew outside of relationships that originated in contractual service.

This does not mean, however, that we should just accept the letters' assertions over those in the book. The letters, too, were written under specific conditions and for specific ends. Readers who enjoy Harrison's descriptions of food will be taken aback by her statement in one of them, "I hate food so much" (FL2). How does this square with the book's keen inventories of her meals on the road, the likes of "Fried eggs with red lard toast stewed navy beans with salt pork onions parsely turnip tops potatoes tomatoes and olive oil oranges" (205)? In this instance I am inclined to accept what the book shows through its cumulative scenes, not hatred but savor.

At the inception of the book project, Harrison had implored Foster, "Please do not ask me any questions about the Book as I never count my chickens before T. a. H." (FL2). The statement demonstrates the importance she attached to authorship, about which she felt anxiously hopeful. And while it also suggests she may have found her friend somewhat importunate, she responded to Foster's sympathetic interest with a full report on its progress.

In perusing the available evidence about Foster, a picture forms of a woman who closely followed not only her friend's world travels and budding authorship but also the activities of other African Americans abroad. Harrison's postscript to the 1936 letter, "am sending some papers that you may enjoy the news of Honolulu" (FL3), indicates that she routinely forwarded her newspaper clippings. The Smith collection includes three such clippings, each pertaining to Black subjects or performance: a story about African American Gold Star mothers and widows in Paris, a notice about Josephine Baker's wayward panther, and a photograph of a group of Spaniards in blackface celebrating the Cabalgata de Reyes Magos. As the recipient of such "news," Foster may have served as a sounding board, helping Harrison assess racial performance overseas, both literal and figurative.

Of course she herself was a writer, too, keeping up her side of the correspondence. Testament to her savvy, Foster navigated the complexity of writing to someone at shifting international locations—calculating postage to different nations, addressing envelopes with place names in various languages, reckoning how to make letters reach *poste restante* locations within fixed windows of time. The sheer fact that Harrison, who was exacting about the mail she received, wrote to her for a good decade proves the quality of Foster's correspondence—the appeal of her sentiment, ideas, and prose.

The *Atlantic Monthly*, not surprisingly, shows no trace of her. There are no letters from Foster in its pages, no shout-outs from editors and reviewers. An enabling friendship between two African American women was well outside the magazine's purview, and even more so, their shared writing project.

The single artifact in the Ann Cunningham Smith collection, the miniature "lap desk" on which Foster wrote, brings her into sharper focus. Scarcely the size of an encyclopedia, the brown leather desk opens to reveal an inkwell, a box of jewel-colored wax sticks for sealing envelopes, and a board on which to write, balanced on, of course, one's lap. We can imagine Foster sitting with this economic, portable alternative to a private study, fountain pen in hand, writing to her distant friend. Perhaps the California bungalow in which she lived had no space for a dedicated workspace, or even for a regular desk. Perhaps she could not spare the time to sequester herself in a single location but rather shifted her writing materials about in concert with home duties. While that lap desk hardly signifies mobility of Harrison's scale, it does indicate Foster's movement within her domestic environment, as well as, perhaps, her family's journey of upward mobility. That her descendants preserved and donated the object shows that her identity as a writer was central to their view of her. Unlike the textual materials, it has no direct connection to Harrison.

Harrison told Foster that she was "glad that you and your Children liked My Great Wide Beautiful World" (FL3). She likely refers to Alice Foster Cunningham and her sister, Edna. In assessments of Harrison's collaborative enterprise, Cunningham can be easy to overlook. Her mother was the recipient of Harrison's letters, and her daughter was their donor. But she too played a key role. For one, in a maternal line of transmission, she likely received the papers from her mother to safeguard, in turn passing them on to Ann.

Lauded on her death as an "LA pioneer," Alice Foster Cunningham is still another putatively ordinary African American woman with a remarkable migratory biography. Born in Texas and raised in California, she graduated in 1918 from Prairie View State College in Fort Worth, Texas, a historically Black college; she also had a teaching certificate from the state of Arizona. She married quite early, a man named Grimes; divorced; and at twenty-two wed David O. Cunningham. At the time she worked as a stenographer at a real estate firm.[63] Eventually she settled into her long career as office manager at Virginia Road School in Los Angeles.

An indication of her status, Cunningham's name crops up in the women's column of the *Los Angeles Sentinel* with respect to her club activities. She was a member of the Girl Friends Club—an association characterized as "a group of socially prominent women," the Sierra Madre Woman's Club, LA Urban League, NAACP, and Outdoor Life and Health Association.[64] Some of them sponsored charity benefits for worthy causes, as when the Girl Friends Club held a dance to raise money for the Woodlawn Branch YWCA.[65]

Cunningham liked to travel. Family photos include one of her and her husband posed in front of a horse-drawn carriage in Victoria, British Columbia. One wonders if she toured Europe, as at the time women schoolteachers made up a significant percentage of American budget travelers on the continent— making such trips, it was alleged, for the sake of their students.[66] Accustomed to both traveling and organizing aid, Cunningham (or possibly her mother) was identified by Matthews as the person "who assisted Miss Harrison in securing her passport when she left Los Angeles."⁶⁷ Passports were once used mainly to keep track of visitors and alien residents within the United States, but after World War I the federal government implemented a more thoroughgoing system to manage the movement of its own citizens overseas. In the 1920s, passport regulations for Americans were still new and confusing, and it is not surprising that Harrison needed help to get one.

For upper-class Americans, who had previously relied on letters attesting to their character, the passport mandate felt insulting. For the lesser heeled, the main objection was the hefty application fee—ten dollars a passport, renewed every two years for the same amount.⁶⁸ This was after all the exact sum Harrison had spent only two years earlier to purchase land in Santa Monica. The Foster family's assistance in "securing her passport" may have included covering the

FIGURE 2. Alice Foster Cunningham (center), office manager at Virginia Road School, Los Angeles. Source: Virginia Road School office, Security Pacific National Bank Collection/ Los Angeles Public Library.

application fee, especially since Harrison indicates that they later sent her ten dollars to renew it in Paris (FL2). This raises the question of what other support they extended from California. Not only the Dickinsons, with their upper-class grand tours, but also the Foster–Cunninghams, with their middle-class trips, offered Harrison travel models and material aid.

The family's status continued to rise. The next generation, Ann Cunningham Smith and David F. Cunningham, evidently entered the upper middle class, with the former a presence in the society pages and the latter a lawyer and judge. "Your Grandchildren have grown so lovely" (FL3), Harrison congratulated Foster. Jefferson recognizes Ann Cunningham Smith as a late informant in the acknowledgements of her book on Black Angelenos' leisure activism. In addition to donating her grandmother's papers to UCLA, Smith contributed family photos to the Los Angeles Public Library's "Shades of L.A." photo drive. They include pictures of the four Cunninghams outside their LA bungalow, her parents on vacation in Victoria, her mother at work in her office, and herself with her children at the beach.[69]

There are also two intimate portraits of her father and her mother, each posed holding an infant David. The baby grew up to become Judge David F. Cunningham, a "Southland leader" and one "of the city's top dignitaries," according to the *Los Angeles Sentinel*.[70] Cunningham was active in Los Angles through the 1970s up until his rather early death in 1984. As a boy, he indirectly structured Harrison's correspondence in that, always short on "pennies" for postage, she postponed writing his grandmother until she was in a new country so as best to augment his stamp collection (FL1). "I will write again when I visit 'Corsica,'" she promised Foster from Nice, "That David may have another stamp for they may use different one there altho it is French" (FL2). The growing collection points to young David's inclination, similar to Harrison's own, to make an imaginative claim to the world.

Still, despite these tantalizing details about her relationship with the Fosters, it is her relationship with the Dickinsons for which the documentary trail is best marked, due to the white family's wealth, status, and stability. For close to half a century Myra resided in the exact same house, at 423 South Lafayette Park Place. For decades Harrison gave that house as her legal residence, and she likely stored her possessions there, given that she told customs officials in Japan that she had left her "baggage" (285) in Los Angeles seven years prior. During her travels, the Dickinsons routinely, it seems, not only forwarded the money she had due her but also made payments on her behalf; in Harrison's words, "keeping up my taxes and looking after my welfare just for Kindness" (FL3). A note she sent Myra from Yangon indicates she sometimes asked them to front her funds: "Write me at Marseilles and if you have a little interest money will you send it if not kindly stick in a love offering I'll keep an account and return it" (150). They sent her the occasional care package, too, including a seven-pound box mailed first class. Not least, Harrison disclosed, each of their letters included "always a little bill 5. 10. 15 or 20. that I may enjoy my self more" (FL3).

This assistance was comprehensive enough to move Harrison to assure Foster that her dependence was only emotional, not financial. "And do not miss understand me for it is only Their richness of heart to heart kindness. for I have lived just as I do now from 1910. and never asked any one for a thin dime" (FL2), she insisted. The relationship certainly invites skepticism, due to not only George Dickinson's paternalism and Harrison's love of perks but also, more fundamentally, the power imbalance between them. Harrison's own statements, however, discourage drawing easy conclusions. In accord with polite convention, she tells Foster, "I am glad your Family are well," only to then immediately identify some more heartfelt sources of gladness: "I am glad I have no one close to me glad I went out from my dozon of Kin before I were old enough to know all their names am sure I could never have trusted either of them my likes and dislikes from a tiny child were always so different. But in Mr & Mrs Dickinson I had an understanding friend" (FL3). Thinking about Foster's family excites relief that she herself has no intimates and sparks memories of her disappointing "kin." Yet she then qualifies the claim of having "no one close," to acknowledge the "understanding" Dickinsons. "They were careful not to let me loose faith in Them," she adds. That she was writing to Foster, as opposed to still another employer turned backer, raises the odds of her sincerity. Offering glimpses of their relationship, her warm comments save the couple from being faceless agents of white paternalism.

As I have argued, by some measures the Dickinsons' contribution to Harrison's world venture has been overstated. Although at its onset, the investments they made for her in their real estate holdings helped fund the journey, by then she was already a savvy traveler with international experience. More important, the segregationist roots of those investments—and of the personal fortune from which their aid flowed—have passed unnoticed, even as the Dickinsons eclipse recognition of other patrons like the Fosters. Yet at the same time, the couple has not been recognized for encouraging her travels and for energizing her correspondence, inspiration that was intertwined.

Ultimately, the letters they sent, which galvanized Harrison's own literary output, may have been their most important service. Harrison stated, "Not another Two would have written me all these years so reagler and such kind encouraging letters as they did" (FL2). None of them are extant, but the chummy, patronizing letter about her character and portfolio that George contributed to the *Atlantic Monthly* surely does not resemble the letters the couple exchanged with Harrison herself. She enthused, "They respected my likes never tried to change my ideas about where nor when to go but just wrote encouraging letters joyful ones and when I arrived in a city town or country expecting a letter unlike any others that would have put it off and it came after I would have gone or not at all and when it did come be full of Ifs" (FL3). Sympathizing with her desire for travel, cheering her on, the Dickinsons could write the kind of expansive letters that captivated her, and they could knowledgeably discuss her destinations because they'd been to most of them, too. Their experience did not make them overbearing. Rather, Harrison emphasized, "They respected my likes."

Harrison's dedication of the book to "Mrs. Myra K. Dickinson" spotlights the role she played in engendering it: "Your great kindness to me have made my traveling much happier if You hadnt been interested in me I never would have tryed to explain my trips also your True and Kindness encourage me and made me more anxious to tell you the way I spent my time." Harrison implies that without the other woman's "kindness" and "interest," she might not have been moved to write. Dickinson's generous receptivity not only made the actual traveling "much happier" but also spurred her to narrate it.[71] When she really had to economize, she spared her stamps for only "my People on Lafayette" (FL1).

Giving more detail about the Dickinsons' influence on her writing, she informed Foster,

> They wrote sweet letters and at each place or Country that I had written to them I would likely be on arriving it would be like getting home as a kind letter would be there to greet me so it made me happy and love the Country and its People and so it made The People love me and thats why their are joy in all of my letters which were sent to a Publisher in hopes that He will also for the joy that they express publish them. The story of my life we have no doubts. (FL2)

The "joy" in her letters, she argues—which induced that male "publisher" to accept them—can be attributed to the Dickinsons. The tone and texture of their writing determined her own. To take this a step further, her narrative persona of the joyous wanderer with her lusty world embrace had its source in the adventurous American couple who gave her a feeling of homecoming in each letter they sent.

In this respect they were the polar opposites of correspondents in her actual natal home of Mississippi, whose dour grudging epistles had distressed her. Maybe it was they who wrote those letters "full of Ifs." Skeptical of her enterprise and suspicious of her motives, just as bad, in her view, they were too local, too absorbed in their immediate social environment. Such a provincial outlook checked any effusive exchange. Her correspondence only bloomed once she began exchanging letters with effusive worldly people. Harrison wished "alway to be where wealth health youth beauty and gayness are" (318), and this preference applied not only to her daily lived experience but also to her attenuated textual one. She told Foster—in what reads almost like a warning to her long-distance pen pal—"I never like to read any hard luck stories in a letter to me" (FL3). Her stated aversion to tales of money woes or other misfortune indicates still another source of her zest for writing the real estate magnates.

To have been sustained all those years, the relationship must of course have been rewarding to both parties. By means of her lively vivid dispatches, Harrison not only sustained but also deepened the Dickinsons' global experience. Their own traveling was now scaled back to trips to resort destinations such as Waikiki, devoid of the risk and discovery that marked their early Latin American voyages and intrepid automobile tours. Visiting many of the same places as they had, Harrison helped them see cherished sights once again. But more important, she gave them access to entirely new types of places and people. Inside a familiar network of major tourist destinations, her travel reports guided the Dickinsons—and, later, the readers of *My Great, Wide, Beautiful World*—through a maze of microsites that would have been alien to them. These include the busy Mumbai apartment complex where she lived among solicitous Indian neighbors, the dance hall she frequented in Madrid, the dining room in a private home in Israel where she slept when no hotel room was to be had, the African "village" at a Paris exhibition whose inhabitants offered her food, and countless others. "Shaking the basemap," to use Judith Madera's metaphor, she took a set of hyperknown locations and configured them anew. Katherine McKittrick states that African American women "have been relegated to the margins of knowledge and have therefore been imagined as outside of the production of space."[72] While the Dickinsons and Harrison's white readership at large may well have suffered from this failure of geographic imagination, through her stream of travel narrative she showed them otherwise, conjuring up novel spaces and places for them to marvel over.

Hinting at the more immediate emotional needs she met for the Dickinsons, one of her letters figures them as "sitting up on the World in Their richness of kindness all by Their Lonsom" (FL2). The by now elderly pair had lost both their infant son and adult daughter, bereavements that might have contributed to their attachment to the gregarious Harrison. Her metaphor, "sitting up on the World" (which she used more than once), suggests that their class elevation further isolated them—the occasional well-heeled vacation aside, in their old age rarely venturing beyond rounds of the segregated golf course, country club, and pricey restaurants that catered to LA's insular upper class. Whereas they were sitting up *on* the world, alone, Harrison was immersed *in* the world—in what she called "my world"—with endless forms of sociability to hand.

After her world trip, writing to Foster from Honolulu, she stated that she was eager to move on. But "never looking towards L.A.," she took pains to add (FL3), having earlier informed her, "I wish nothing from Los Angeles." There are many reasons why her friends might have expected her to "look towards" that city. In Los Angeles, she had forged the mentoring friendships with Black and white Angelenos that she identified as her life's most important, made the investments that launched her world travels, and developed the connections that later made the city a center for the promotion of her book. She had even bought land. While traveling, moreover, she identified herself as "from Los Angeles" (9), and she always gave the Dickinsons' LaFayette Park home as her US address. Nevertheless, a quarter century would pass before she returned to California.

PART II

During *My Great, Wide, Beautiful World*

World Travel and Authorship

CHAPTER 4

Writing and Traveling for Freedom

On a "beautiful June morning," Harrison stepped aboard the German ship SS *Ventana*, feeling "happy," in her words, "that I had no one to cry for me" (1). New York City was at peak cultural ferment, but she chose to leave it behind for England and beyond. Eight years and more than thirty countries later, as Honolulu hove into sight at the close of her around-the-world journey, she endorsed her solitary mode once again, exclaiming, "How well we can carry out our plannes if its just yourself" (310–11). Yet although her travels were resolutely solo, through prolific correspondence she directed a wide, continually growing social network. Letters fostered her preferred way of conducting relationships—at a chary distance—and a selection of them comprise *My Great, Wide, Beautiful World*.

The three chapters that make up part II of this study, During *My Great, Wide, Beautiful World*: World Travel and Authorship, chart Harrison's activities between June 1927 and April 1935. The majority of this roving history is gleaned from her book. Two of the chapters include critical readings of *My Great, Wide, Beautiful World* itself, folding textual analysis within thick description of Harrison's lived experience. Although the text is overdetermined—crammed with acts and scenes that support multiple interpretations that in turn shift when larger contexts are supplied—I resist the temptation to write a tome like Harriet Beecher Stowe's *A Key to Uncle Tom's Cabin*, a concordance heftier than the text it glosses. Rather than explicate all the book's events, I highlight key ones, especially those that took place in Paris, Madrid, Djibouti, Mumbai, Cairo, and on the Cap d'Antibes. In respect to biography, Shirley Leckie states, "If we have done our work well, our biographies will not be 'definitive,' in the sense of defining a person once and for all. Instead, our greatest accomplishment will be to present a portrait of that individual that will motivate others to conduct their own research."[1] I bear in mind her caution, that the goal is to open up a subject rather than close it down, with respect to my analysis of not only Harrison's unusual life but also her unexpected travelogue. The brief readings proffered here can only intimate all that this text, immeasurably rich, offers.

This chapter, "Writing and Traveling for Freedom," investigates how Harrison's expedient literacy and astonishing mobility—which were mutually constitutive—enabled her to satisfy what she called her "main point" (315), freedom. Such freedom was not only spatial but also subjective and societal. By virtue of living in and moving through numerous places—and by writing about them—she sowed an abiding cosmopolitanism, inhabiting diverse subject positions and entering diverse social relationships. With its controlling themes of literacy and mobility, *My Great, Wide, Beautiful World* updates a long African American life writing tradition that likewise features them. The first question usually posed about Harrison is, Why was she so avid to travel? But we can also ask, why was she so keen to write?

Shortly after the journey began, she reported on a conversation with curious fellow travelers in Scotland: "They say are you going to have a holiday I said yes for about 2 or 3 years" (9). Ultimately that "holiday" was three times that long. It can be divided into an initial three years of work and travel in Europe, the Middle East, the Indian subcontinent, and Europe again; an extended stationary period in France; and a final eastward journey to and through Asia.

During the first months of the venture, Harrison held housekeeping jobs in London and Paris. In Paris she met the Morris family, who in time would cultivate her authorship, but she worked for them only briefly before moving on to Nice, long a magnet for American expatriates. From there she launched an intensive fourteen-month tour of Europe and the Middle East, with India her final destination. She worked in Cairo, lingered in Sri Lanka, and on reaching India toured the country and took jobs in Mumbai.

She had weighed concluding her journey there and returning to the United States, but instead she doubled back to Europe. While she explained, "I did not want to finish up my around the world trip from India as I wanted to be in Europe for the Passion Play" (165), clearly her motives ran deeper than theater. She worked in several cities in Spain and then in Paris again, having been away from the French capital for two years. Additional travel in northern Europe brought her to September 1930, shortly after which she began a close to four-year residence on the French Riviera. As her editors explained, given the political conflicts in China, it was not a good time to continue "her journey around the world" (265) by traveling east to that nation, as she had once intended. For Harrison personally, moreover, active travel anywhere had become more challenging, since the October 1929 stock market crash wiped out the income from her Los Angeles real estate investments, on which she had relied. This long interval also included more time in Paris, during which she worked on her book manuscript with Mildred Morris. Finally she executed a rapid eight months of travel: around Scandinavia, across Russia, and in China, Korea, Japan, Taiwan, and the Philippines. From Manila she launched her return voyage to US territory, stopping off at several Asian ports before reaching Honolulu.

A casual reading of *My Great, Wide, Beautiful World* might support the belief that Harrison drifted from place to place, especially since she often stayed much longer at a destination than planned. Yet she crafted her itineraries quite deliberately. Consider her workaday but revealing remark about departing southern Spain: "I hate to face the cold at Madrid but I have spent what I had planned to spend here. and expect to leave Sevilla for Madrid at 8.40 A.M. Feb. 6" (208). Climate was always a consideration, and she increasingly sought out the sun. Alternating stretches of leisure with longer stints of work, she meticulously managed her finances, and her book demonstrates her tactics for acquiring the necessary tickets, visas, and currencies. In respect to exchanging the latter, she boasted, she had it "down to a fine point" (31). Her keeping her correspondents apprised of *poste restante* locations is only one sign of her scrupulous method.

Its length aside, her travel program was a conventional one, studded with standard tourist attractions. In England, she toured Shakespeare's house. In India, she saw the Taj Mahal; in Egypt, the Great Pyramid; and in Jerusalem, the Western Wall. Having pored over guidebooks to Rome, prior to leaving that city she insisted—twice—"I had seen everything" (34, 35). During her visits, she was well served by developed tourist infrastructures, availing herself of American Express and Thomas Cook offices, Baedeker directories and travel magazines, tours and excursions, guesthouses and YWCAs. She plotted her route to coincide with major spectacles, such as the Manila Carnival in the Philippines, the Barcelona International Exposition in Spain, the Oberammergau Passion Play in Germany, and the cherry blossom season in Japan. She had no desire to leave the beaten track, despite occasionally straying, as when she worked in a Catalan village or crossed through present-day Bangladesh. Yet even as she presents a familiar set of tourist destinations, she maps a countercartography of the alternate places, events, and human interactions she experienced at them—which is what makes her narrative so gripping.

The published book consists of over two hundred entries, each headed with place and date. While the majority of them had an earlier rhetorical life as personal letters, collectively they were made to appear like a travel journal. The salutations, valedictions, and personal addresses to their first readers were excised, in what amounts to considerable editorial intervention. Nevertheless, sentence by sentence, Harrison's original prose is preserved. Her text is visually striking for its nonstandard grammar, phonetic spelling, absent or extraneous punctuation, and uneven capitalization. Morris expended considerable painstaking effort to accurately transcribe Harrison's handwritten manuscript.

The idiosyncratic orthography belies the creative richness of her writing. Or rather, it helps generate that richness, a relationship of which many of her readers were well aware. They include the *Des Moines Register* reviewer who maintained, "Her erratic spelling and grammar only give her rainbow writing a deeper hue."[2] Harrison herself knew that the tactile distinctiveness of her writing made for its power. In response to the suggestion that she publish her letters "just as I have written them misteakes and all," she speculated on the outcome of a different choice: "I said if the mistekes are left out there'll be only blank" (243). The joking statement may indicate real alarm about the prospect of invasive editing that would render her prose "blank," erased.

Such trespass would threaten not only her writing but also her selfhood, given that her language was bound up with her regional and racial history and identity. Within *My Great, Wide, Beautiful World*, her community of origins is on display in the composition of her sentences even as it is not explicitly referenced. Elements of her prose, according to one linguist's informal assessment, are both distinctly regional and evidently African American. Chimene Jackson characterizes it as "the native, naïve vernacular of a Black woman," patently "sopped with memories of a Southern upbringing."[3] Harrison succeeds in replicating natural speech patterns, capturing her idiolect on the page.

Her fused sentences often build toward surprise endings. Consider, for example, her description of the view from Darjeeling: "Up and down are the thousand of light as this is a City above and below mountains and clouds the lights seem to mix with the stars" (140). In another India scene, she calls attention to her language dexterity in transcribing a conversation about her response to the Taj Mahal. "The night was getting dark the dew was falling heavy and I said, 'I would just like to put a glass over it I feel I must cover it over,'" she relays. Her companion replied, "That's beautiful" (133). Even as she states "it cannot be described eyes must see it" (134), she suggests that her "beautiful" words compensate for the sight. The book sustains this pungent immediacy throughout.

The effect can be overwhelming. Well expressing the challenge *My Great, Wide, Beautiful World* poses, a Canadian reviewer sagely advised readers to "dip into it at intervals, as its 318 pages of queer phraseology and minute detail will produce a mental surfeit if attempted in a single feast."[4] What one registers shifts with each reading, and it is surprisingly easy to overlook key scenes or even misapprehend full passages.

As a whole, moreover, it can give a misleading impression of Harrison's experience. Given that the book is a hectic blur of travel narrative, one might readily conceive its author's life as a hectic whirl of travel activity. In reality, Harrison's movements were far from frenetic. She was stationary and employed for close to twenty-three months of the initial three-year travel period that comprises the majority of the text—that is, for approximately 60 percent of it. In my view, the astounding swiftness of her enterprise is located not in the rapidity with which she passed through different places but rather in the rapidity with which she cycled through different jobs, a feat that has received far less comment. Between 1927 and 1930 alone, she held thirteen paid positions, including that of maid, nurse, companion, housekeeper, and cook. Their tenure ranged in length from three and a half months, as a nursemaid in Cairo, to scant hours as an "upstairs girl" (201) in Seville. If we calculate the ratio of her work time to her travel time, labor trumps leisure by a clear margin. Yet readers might register the ratio as the reverse, since the text devotes far more pages to her leisure and touring—measured by the day—than to her periods of work and residence, which are usually sketched in broader strokes, presented in blocks of weeks or months.

Nevertheless, Harrison's attention to work is steadfast. The opening entry of *My Great, Wide, Beautiful World* inventories the contents of her suitcase: "2 blue dresses 2 white dresses and one black aprons caps and references" (1). That it should be stuffed with work clothes—and "references"—forecasts the book's controlling interest in getting, holding, and leaving jobs. Once in London, Harrison shows herself assessing the city as both a tourist and a prospective maid: "The first look at London I liked it. It was a beautiful morning the maids in their neat blue dresses and white caps was cleaning the brass whiting the door steps and scrubing the sidewalks. I promised myself when I Looked For a job not to get in a house with such work but instead a appartment" (4). The phrase "I promised myself" marks the slippage from the tourist gaze to that of the vigilant future servant, and this double perspective is evident throughout the text. She shifts from one to another and even combines them, as when she configures the job hunt as a way to explore a new city. Indeed her word for a destination and word for a job are one and the same, "place."

The book delineates her methods for finding work on the road. She habitually first chose a desirable location and only on arrival searched for a job, variously relying on employment agencies, newspaper advertisements, bulletin boards, and word of mouth. A report from Barcelona exhibits her routine practice. "After looking over this nice little city," she explains, "I felt I would like to stay a while. So I looked up the news Paper Offices and this afternoon will have an add put in" (171). She continues, "I have just found the number of an employment office in the paper I picked up at the Bull ring so will first go there before I put out an add. as I can learn much by listen to the girls talk. I hope to get something as I would like to stay here" (173). Appearing in English-language papers, her ads for her services were succinct notices like the one she placed in *The Times of India* under "Situations wanted": "Experienced woman desires place as Nurse or Housework" (114). As her experience accrued and her professional network broadened, she reached a level where she could pick and choose among jobs. She boasts, "I dont know much but I'll always know how to get a job I dont care about what country I am in" (218).

Demonstrating her ability to land prime positions, many of her employers moved in high places, including an Italian aristocrat with friends in the English Parliament and the former Broadway star who became her editor in Paris. Harrison gravitated to employers who were not native to the nations in which they lived: English women in the United States, Cuba, Canada, and Egypt; Bolivian and Italian women in England; English and American women in France and Spain. "The foreign people are the best to work for" (12), she affirms in England, without specifying any one nationality. Expatriate households made for the pleasurable sensation of living in two countries at once. Of more material import, they were less likely to enforce customary workplace rules and more likely to pay above the going rate. Of an American family in Madrid, for example, Harrison notes that they covered the rent for her private room and paid her "5 times more" (211) than would the Spanish. The attraction to "foreigners" worked both ways; part of her appeal to that particular family was her ability to translate their needs from English into Spanish and French. She coupled her personal enterprise with the standard immigrant practice of importing skills to places where they were in short supply.

While expatriates of a number of nationalities were drawn to her, as a fellow outsider who could mediate between local ways and their own, American employers were a clear majority. We can speculate that for some of them, the sheer fact of Harrison's race contributed to her desirability, in that it broadcast her role in their household similar to how it would in the United States. In oral testimony collected by Judith Rollins in *Between Women: Domestics and Their Employers*, Odette Harris, a former maid, reflects on the source of an employer's preference: "She used to say I was more intelligent than her previous Irish girls. But I thought that the advantage of having a black girl is that it shows automatically that they have a maid. If someone comes to the house and sees a black girl, they know she's a maid.... We give them much more status."[5] Even when physically outside of the United States, Harrison could find herself within an American racial economy. She nonetheless enjoyed far more ease and flexibility in her work arrangements overseas than was conceivable in the American South. She was so competent and winning that her employers not only offered material inducements to get her to stay on but also allowed her to challenge their mandates.

Through both its content and its epistolary history, *My Great, Wide, Beautiful World* inscribes her relationships with her employers, virtually all white women: some friendly, some hostile, some more intimate than she wished. From the village of Molina del Rey, she reports, "The Mother said that everything I know and do it just perfect and if I stay all things she will give me and that it seem I have been here a year because she never tell me anything. I thought to myself thats about the nearest to a year I'll ever get" (183). She recalls of an invalid employer in Mumbai, "She would call out from the bed 'Oh, Frank you are not there talking to that young woman I wont stand anything like that'—and she would throw pillers. then I threw pillers and she would say 'Come here, Frank, I'm afraid of her She has a temper,' and he would say, 'Well so have you a temper'" (120–21). And in Madrid: "I stoped working for Two days we had a little falling out then I came back and got more money and more time so we are very happy and I am some lucky dog" (212).

Harrison writes, often critically, about the people who hired her in a book mostly composed of letters written to other people who had once done the same, and subsequently redirected to a middle-class readership at large. The text thus becomes a kind of trainer's manual, educating employers about acceptable working conditions for their servants. It warns them, moreover, that one's maid is watching you and perhaps even writing about you. Jackson identifies the racial dimension of her critique: "Juanita's entries embarrass the hubris of the white [colonist] expat, European rover, and frail white women who have many possessions but only one language, and maneuver countries on the basis of their white privilege.... Her freelancing exposes whites' servitude to being served."[6]

Yet she had to be circumspect, her letters only implicitly criticizing people akin to their recipients. Even from a textually mediated distance, Harrison expressed her views with care.

While she focuses less closely on them, she offers glimpses of other servants, too. *My Great, Wide, Beautiful World* files an array of snapshots of working people: Spanish maids, Filipino sailors, Chinese cooks, the African American nurse who moved with her employers from Albany, New York, to Rome, the Indian boy who offered to clean her room for free, the colonized groups exhibited at fairgrounds. She indicates both these workers' proud resourcefulness and their exploitation at others' hands. Like her, many are far from home, but save one, none move freely from city to city and job to job. She presents even that exception as constrained, a Burmese woman she met in Kolkata who "are working her way about India but are afraid to take a chance on going to Europe" (139).

Harrison countermands a prevalent belief in her day, that African American women must always work. Not only does she quit jobs in order to move to other places, but she also sometimes quits them simply to be idle in the same place. "Nothing are so nice as loafing when you want to" (86) she stated on leaving a job in Cairo, and on doing the same in Madrid, "it is good to be a loafer again" (219). Her relish of what she pointedly identifies as "loafing" conjures more than just the onerous labor of her youth. It also evokes efforts in the early twentieth-century South to criminalize African Americans who would control their own movement and time, efforts codified, to quote Risa Goluboff, in a "rash" of laws that targeted "habitual loafers."[7] Goluboff argues that although vagrancy laws were ubiquitous nationwide, their purpose in the South was "to return black Americans to a state as close to slavery as legally and practicably possible."[8] She explains, "Whether African Americans tried to challenge their prescribed place by moving farms, changing jobs, frequenting public places, or choosing sex partners, vagrancy laws were always an option."[9] The Georgia nurse who testified in *The Independent* succinctly summarizes how these measures shut down bargaining power: "We have to work for little or nothing or become vagrants!"[10]

Escaping such control, Harrison fully exercised her right not to work in a myriad of international locations. In respect to taking the waters at Baden-Baden and availing herself of hotel room service, she remarks, "I am the real lady now. I worked 3 months so now I must enjoy it in a beautiful way then I will be willing to work again" (257). Her avidity for attentive service speaks to her long past and present reality of domestic service. Even as refraining from work makes her not a vagrant but a "real lady," she savored her leisure time in a different way and for different reasons than would a wealthy white woman. Judith Madera identifies this manner of "work stoppage" as a form of "Black freedom."[11]

Transience, manifestly, was Harrison's method of choice to amplify her freedom. To land a job she never hesitated to lie about her timeline, and when her restlessness swelled she left even positions she liked. Along with her wanderlust, she may have been motivated by the simple fact that no matter how good a job she had, it was always a service position, by definition subservient. Thus for all her pride and satisfaction in the deft execution of her duties, in the benefits enjoyed and rewarding relationships formed, no workplace could keep her. She confides, "I always get a job when I go out to get one but never feel any to glad no matter how good it is. Its when I am ready to give it up that I have the grand feeling" (253).

Her book maps the tremendous amount of ground—and water—she covered. Regularly venturing into forbidden territory and arrogating public venues for her own uses, she commanded the space she moved through. Page 1 sets the stage with an account of her first impulse on embarking in Hoboken for England: "Our cabins looked good. I always want an upper berth I dont want anybody making it down on me. I went to the 1st and 2nd Class. Their towels looked more linnen so I took two, the soap smelt sweeter so I took 2 cakes. I went up to the writing room and the paper was the kind you love to touch so I took much and tuked it away in my bunk" (1). At the moment her world journey commences, Harrison literally elevates her status through this series of appropriative acts: claiming the top bunk in her third-class cabin, taking the superior soap and towels from "1st and 2nd Class," and finally ascending to "the writing room" to pocket the fine paper she figures as near seductive. The last act, of course, augurs the book itself.

Throughout her world tour, which was bookended by transatlantic and transpacific voyages with a pod of shorter ones in between, she relied on ocean travel. In traveling thus she participated in a middle-class leisure practice that burgeoned in the 1920s, an unplanned outcome of new laws that slashed the number of immigrants who could enter the United States. To stave off "devastating" financial losses, steamship companies converted steerage quarters into budget tourist cabins, making sea voyages much more accessible to ordinary citizens.[12]

Gary Y. Okihiro reminds us that oceans "are extensions of lived, worked, and imagined spaces."[13] While accounts of Harrison's experience can give short shrift to her time on the water, as mere transit, these interludes were potent.

Twenty-three of the book's entries were written while she was at sea, chronicling roughly fifteen weeks in all. They show that on board she enjoyed communal meals, attended religious services, visited the library, and laid travel plans. She relished the unique social world of each ship, built from the flag it bore, the passengers it carried, and the workers it employed. "On the High Sea are like on land this Steamer is a perfet home to me" (309–10), she enthused en route from Kobe to Honolulu.

At sea Harrison could experience the character of a nation without feeling subsumed by it. With their international assemblies, moreover, many ships offered simultaneous exposure to more than one culture—her favorite environment. Embarking on the SS *Chantilly* in India, she states not only that "its good to be with the French again" but also "the Waiters and cabin Boys are Chinese which I like" (157). But most important, she depicts ocean voyages as a platform for building Black subjectivity, not destroying it, as was historically the case throughout the Atlantic slave trade. In *Slavery at Sea: Terror, Sex, and Sickness in the Middle Passage*, Sowande Mustakeem shows how slave ships "served as vital sites of power sailors used to dehumanize captives, enforce dependency, inflict pain, establish authority, and prohibit any sense of control over one's personal life."[14] While these trade vessels of the eighteenth and nineteenth centuries were the location of "cyclical assaults on slaves' personhood,"[15] Harrison recruited the ocean liners of the twentieth century to fortify hers. A water metaphor, unlike the earth metaphor of roots, well represents her life endeavor, thinly dispersed and spreading everywhere.

During these voyages, as her book would have it, she got along swimmingly with all. Yet surely, her racial identity influenced how some of her fellow passengers interacted with her, especially the Americans. In a 1936 account of his trip to Central and South America, written for the African American readership of Los Angeles's *California Eagle*, NAACP field secretary William Pickens describes the kind of guarded behaviors Harrison might have met with at sea:

> This is the day for my fellow passengers, white, to get tame.... Always on the third day, if you have been absolutely, but unostentatiously unconscious of their existence, they begin to get worried about it and want to find out just who you are, where you are from, what you do, and what you are like—and they look you up. In fact, it began last night, first two or three men sidled up to me: "Taking a vacation?" And one thing led to another—and I could hardly be rid of them.

Continuing his narrative of these overtures, Pickens suggests that Harrison's book may have given him an agreeable sense of companionship with another Black traveler: "And today, bless you, after we returned from the shore visit, the gang came, crowding around my deck chair, where I was trying to get some more fun out of Wood's 'Heavenly Discourse' and Juanita Harrison's 'Great Wide Beautiful World.'" He would like to read in peace, Pickens implies, but "the gang" is intent on descrying why this African American man is traveling.[16] Harrison does not describe any real confrontations, but she does show her efforts to secure privileges and defend rights—taking the better bed in her cabin, informing greedy tablemates that the dishes were meant to be shared. She believed she merited a comfortable voyage regardless of who she was or the fare she paid.

A similar emphasis on dictating the conditions of her international rail travel evokes past constriction at home. Patricia Yaeger identifies "the Jim Crow South" as "one ground against which the figure of her journey gains meaning," and Harrison's Mississippi past manifests itself in her routine travel behavior.[17] She was avid to requisition greater privileges, as when in Europe she boarded trains on low-fare tickets but then proceeded to "go forward to the front" (50). Conversely, in Asia she refused to buy the pricey first-class fares expected of her as a Western woman, insisting on her right to travel second and third class. Her resistance to travel hierarchies is on display in her anecdote about the "plenty of high class Indian women" who travel third class by rail: "They dont want to mix with the Europeans or with classes—then they say to a low cast on a bench you get down there and lay down—when I didnt get down she sat and looked at me with a puzzled face" (110). Owen Walsh states that despite Harrison's disinclination to generate "political critique" from the incident, "its cross-cultural resonance . . . give[s] it a special power and significance beyond mere personal remembrance."[18]

Her resolute claim to physical spaces finds further expression in her domestic practice, a subject I treat more fully in an earlier study, *Playing House in the American West: Western Women's Life Narratives, 1839–1987*. Just as Harrison shifted between the identities of servant and tourist, she shifted between those of traveler and housekeeper. Travel affords a legion of domestic opportunities, as one repeatedly establishes a domicile, for days, weeks, or months at a time. Harrison's paid employment, furthermore, consisted of servicing other people's households. The paradoxical result is that her travel book can read like a meditation on homemaking.

This meditation, however, dislodges the usual class, race, and gender meanings of domestic acts. Harrison spotlights a series of protean homes for one. To take one example, in respect to the Spartan apartment she rented in Mumbai, she comments,

> I think it just wonderful to live like this, my stretcher cost $1.80 which is 5 rupees, a old rose spread which answers for sheet at night cost 36 cents. a zink water pail for 12 cents. a yard and half of pink cotton crape for the Hall window 12 cents. My coat and underware make a good Pillow the suit case made a nice table for my Bible and clock. I have a thick piece of brown paper with a nice handkerchief on it where I keep the 7 or 8 good books I get from the Y. (118)

Her depictions of such dwelling places, arranged to meet her fastidious needs, undercut both views of domesticity as exclusively a prop of patriarchal households and views of African American women as exclusively workers in other people's homes. Yet all the same, this periodic revelry in the domestic never stanched her determination that it not impede her. She invested as little as possible in her short-term residences, and she absconded in the middle of the night from that Mumbai apartment rather than give the required thirty days' notice.

Harrison also had a proclivity for sleeping outdoors, which in light of the vagrancy laws of her native Mississippi takes on larger meaning than simple preference. In Lucerne, Switzerland, she resolved to pass a night on a riverbank, recounting, "It was a Geliouris night I streched out and thought why should I pay to be in a room and shut off from all this beauty" (43). "We would call the bridge," she reassured a concerned passerby, "my Parlor" (45). The habit persisted in Honolulu, attracting the attention of more than one journalist. Sleeping outside in public can be seen as the ultimate act of vagrancy, defying authorities to charge one. With the abandoning of all defenses, it is also an expression of supreme physical security.

Whereas in the United States African Americans were at heightened risk of arrest for perceived infractions, overseas Harrison depicts the police—both the arm and the symbol of state power—as not only supportive but also impressionable. Her account of an early morning arrival in Baden-Baden includes this remarkable encounter:

> It was a beautiful dewy morning just getting light.... the air was sweet and Brisky. while roveing about the Streets Two night watchmen and a Policeman with a police dog came to me and while they were coming I made up my mind to get arrested just to see the jail and maybe a hot cup of coffee. they ask me what Hotel I was stoping at I said no hotel.

if I had no money I said yes. they ask me to show it I said I would not. if I had a passport I wouldnt answer yes or no. I was speaking all the time in English French and Spanish and they were trying to understand what I was saying. after all my durings they didnt make any move like taking me to jail so I took out my passport and let them see it and we parted the best of friends. (255)

Throughout the interaction, she displays a stunning assurance of personal impunity, not to mention linguistic dexterity. The police are like any other men, susceptible to her charm. Of the "adorble" officers she encountered in Paris, with some surprise she reports that "although they . . . keep order they can just be human" (17).

The direct address to her readers in the book's epilogue, which rounds off her description of her unorthodox life in a tent in Hawaii, spotlights her resistance to conventional domesticity: "Well you have bring out your moth ball smelling cloths and no doubt feel very pleased with the world to be in a caged up Building looking out on others more caged up. I have gone through the same and how greatful I am to myself" (318). Indicting her readers for their capitulation to bourgeois convention, arguably Harrison felt free to make this pointed accusation because the epilogue, unlike the majority of the book, did not originate in personal letters. Conjuring moldering possessions and specious ritual, smacking of mothballs, she renders permanence unsavory. Note too that with the cryptic reference to a past confinement, she frames her travel achievement through not only her readers' domestic captivity but also her own: "I have gone through the same." When and where—and perhaps with whom—had she been stuck? Was she thinking of those "years in N.Y." when she was "a love of the Army" (102)?

Harrison is explicit that her freedom rests on her solitude, and the remark about the bygone "love" constitutes her sole known allusion to a romantic partner. Rather than further commentary on her relationship with him or anyone else, throughout the book she foregrounds her status as a single woman. Early on in her journey, she stated, "I no longer wish that I was a man because I can travel with even less than they" (40–41). That she once fantasized about being a man is a tantalizing disclosure: We can speculate as to whether she wished it only to ease her travels or whether it was to ameliorate other aspects of her life. She was soon tutored in the advantages of being a solo woman on the road. The benefits stemmed from both the piquant companionship and material benefits she received from male admirers and the warm friendship and cultural access she gained from female associates. Nonetheless, she was chary of getting close to either sex. Men potentially compromised her reputation and hindered her free intercourse with other people, while women could make too many personal demands.

Harrison's looks, charm, and apparent friendlessness prompted countless useful services from gallant men of all nations and stations. More than once she ironically calls herself a "poor lonely woman" (282), allowing "that lonely look helps me a great deal." The text demonstrates her avidity to take whatever was on offer, whether seats, rooms, meals, or tours, and then with speedy dispatch giving her suitors the shake. "I am auful foxie" (46) she boasts of her ghosting techniques. She had to finesse how to get what a man had to give without letting his presence scare off other benefactors. Regarding the "Italian Proffesor" in Rome who tracked her down at a favorite restaurant, she gripes, "I never let any man go to the place I eat and so I were really sorry when I saw him because I had been going so quietly there and they took more interest thinking I were alone. I thought Now this is going to spoil everything" (34). As some compensation, she adds, "He took me to hear Aida" (35).

Her aloofness was dictated in part by expediency: She was concerned that conspicuous male company would shut down the special privileges she enjoyed as a lone woman traveler. However, she also reveals a deeper fear, rooted in her youth, that it would compromise the upright reputation she was intent on preserving. As a girl, a seeming scandal had caused her to feel wary of her "kin" and dismayed by her "race" (FL2). "I have been careful from an early age to keep my life clean" (FL2), she stated, and she explained that the reason she always carried a "big Bible" was because "my Bible ... speak for me" (112).

None of the women in her book share her talent for turning their gender to their advantage. On the contrary, being female seems only to check their independence, agency, and sense of adventure. Harrison frequently expresses pride in her singularity. Of her accommodation in Aswan, Egypt, she comments, "I am the only Women in this Hotel which pleases me" (92). "No other woman are up here" (308) she notes from the top deck of a Filipino boat; at a dance hall in Madrid, she is "often the only woman except the hired Girls" (219). (She visited the hall "just to see what go on," she claims.) Other women, lacking her zest to explore, show little curiosity about the world outside their home.

Thus she states, "I always feel a little Boy can help a stranger more than anybody else I have found it so in many places. But a woman or a Girl I never expect anything they never know" (259). Moreover, among boys, she alleges, "that's where I am at home as they like myself are natural at all times" (303). Maybe so ... but she is patently uninterested in describing them. Nor does her book offer a single portrait of a boy who resembles her. It does, however, depict two little girls with whom she has much in common, at each end of the class spectrum—a spoiled birthday party guest and a lone rogue living off her wits. She describes the former as "Alice American Born in Habana she speak Spanish English and French and dance the Spanish Dance graceful. But are a swift little lier" (216). More serious and more poignant is her portrayal of a "little unkept Indian Girl about 7" whom she met on a train in Sri Lanka. At first she thought the child was with the Tamil women she hovered by, but she soon discovered she was a "wise little Hobo" traveling without a ticket. She recounts, "She took a fancy to me after some time I took out some change to count it and she begin to beg me for some. I offarred to buy her pop but she refused after we had traveled a while and she knew I wouldnt give her money she took from around her waist a little white bag and before my eyes begain to count all the time looking at me sideways Then we had a good laugh" (106). The wily child and the wily woman are kindred spirits.

While Harrison's interactions with men are zesty but dispassionate, those with women are far more invested—often fleeting, but potent. Even as she insists on her difference from other women, she seems to care quite a bit about them. Other women trigger feelings of superiority, obligation, and empathy, and her attitude toward them exhibits a host of contradictions: avoidance and attraction, solidarity and scorn, kinship and distance.

Her tone can be dismissive. She insists, "I never had a traveling Companion and If I just had to have one it would be a man and of all things not a Woman" (104–5). Just as the men in her text consistently appear as benefactors, bestowing upon her presents, concessions, and aid, the women consistently appear as supplicants, extracting from her many boons. The text displays her persistent impulse to perform small acts of sisterly charity, such as giving her share of a boardinghouse dinner to a "young, blonde dancer" who "looked so hungry" (4) or donating an unneeded coat to the Africans she met in Paris. She finds women's physical needs burdensome enough. "I never go with a Lady because you must pay her carfare they like to stop and have a cup of tea another stop in the Publice House for a glass of beer and another at the W.C. and the time have pass and your little change" (6), she remarks, amusingly. But more onerous is the weight of their emotional needs. She often gets drawn in despite herself, as when she helps a woman in Syria get her daughter onto a train just as its doors close: "Well the mother cried and went on it made me cry" (68). After a friend's mother died, she spent some weeks at her house in a dark cheerless neighborhood of old Cairo. "I wanted to be at the Y.," she explains, "but felt sorry for Her so I went to stay" (83).

Harrison also highlights the unappealing lives of married women, confined to their homes and hard at work. Of villagers in Spain, she recalls, "I watched the women washing cloths and had great fun with them. I told the women and girls if I was married I would be at home washing dirty cloths just as they are during" (192). Likewise: "I teased Josefa the Young Wife she bought rabbit I said you have a husband so must eat rabbit I am single so can have turkey" (206). Summing up her outlook, she rather harshly states, "I have not patience with married people as a rule they make such a mess of their lives" (209).

Her self-portrait is textured by numerous comparisons, always to her credit, between her practice and that of other women, whether they be single, married, traveling, or employed. So many of the women she meets offer negative models, showing her how not to live. Yet her emphasis on their lack of agency registers a pressing awareness of how economic factors, social mandates, and internalized gender beliefs constrict them. Her often unsympathetic, yet steadfast, attention to female powerlessness makes for a form of solidarity. In suggesting that the only way for a woman to have a fulfilling life was to be as exceptional as she, the text verges on patriarchal critique.

"We have the freedom to make [life] whatever [we] wish" (FL3), Harrison assured her friend Alice Foster, a conviction that made her impatient with those who bemoaned their lot. Yet it is evident that much of her autonomy overseas depended as much on her American citizenship as on her personal intrepidness. The discrimination she faced in her home country notwithstanding, as an American she could carry a strong passport, make influential friends, invest her savings, and spurn marriage. Partially funded by real estate investments in Los Angeles and facilitated by friends in that city, her experience abroad was routed through the United States in quite material ways. Even more so, of course, her book: mostly composed of letters mailed to and retrieved from the United States, edited by an American woman, excerpted in an American magazine, and published by an American press. Mrs. Morris, her former employer in Paris, enclosed a "new dollar bill" (166) in each letter she sent her.

Harrison pointedly distances herself from her nationality nonetheless. Her book suggests that to be American is to be indulgent, profligate, and culturally obtuse. It also posits the default American identity as white and middle or upper class. Thus, she confounded Japanese customs officials on her arrival in Kobe: "They was looking for the lone American and didnt know when they saw me" (285). She startled the ordinary Japanese citizens she met, too. "The real shock," she remarks, "is when They learn from my Passport that I am an American" (286).

Harrison's criticism of other Americans is usually directed at wealthy individuals, like the "everlasting hansom American all brown from the sea and sun" (27) in Antibes who pawned off others mens paintings as his own, or the provincial employer in Madrid dismayed by the lack of American goods. Broader national critiques are routed through the perspective of foreigners. For example, she quotes an Italian's indictment of US immigration policy: "He said if Columbus come back now He would be sorry He discourved America when they refuse to let them come in" (31). More severely, in reference to Katherine Mayo's sensationalist attack on the Indian independence movement, *Mother India*, she relays, "a Hindu said that he would go to America and write one of all the Black sides of America and call it Mother America" (156). Her response? "I told him thats a good idea" (156–57). Note the "color-coded wording of the critique," to use Walsh's phrase.[19] While Harrison does not herself condemn the nation, she quotes those who do. Symbolically enough, she disembarked in England on July 4, a date that her German ocean liner had commemorated with a special meal in a nod to its American passengers. Harrison secured her own American independence by leaving the United States.

Yet for all this, the book includes many expressions of national affiliation. After all, she enjoyed celebrating the Fourth of July on that ship, with satisfaction observing that "everyone seem to turn American the table all in stars and strips Flags" (3). She refers to "my country" and uses the pronoun "our," and numerous textual moments prove her sense of connection. American authors wrote the best books, she maintains (118). She notes when she meets someone who had lived in one of the same American cities as she had. When she sees "Ringens Bro's Greatest American Circus" outside the train station in Tours, she "knew that it was a lovely place" (237). She misses American Christmas. She is moved by the Suresnes American Cemetery. Of a job with an American family, she remarks, "I get a big kick out of it its just like a visit to the States" (249–50). She delights in an uncouth group at a bullfight, "Two young Handsome American Couples" who were "all chewing gum it looked very funny to see them chewing but I got a kick out of it being so home like" (229). And there are moments when she simply sounds proud: "Across the Hall are a Parsee Young man also a Student He say to me the English is nice but oh. I get so chilly when they come near—then he say youre not English are you? I told him No. I am american" (119).

While she oscillates between rejection and gratification in respect to her nationality, she discloses no ambivalence about her identity as a Californian. As discussed in the previous chapter, this she simply embraces. Los Angeles was her long-term city of residence shortly before she left the United States, and she resolved to submit California as her home state while abroad "as every one know it" (192). The people she meets, she imparts, "think it is so wonderful I am from far away California" (46). Her association with the Golden State made her seem glamorous, and in a more profound way, even as she decried the superficiality of Los Angeles, its hybrid, innovative character made it an apt home for her.

Regressive, static, and poor, the South's reputation was the antithesis of California's. It is no wonder that Harrison did not identify herself as a Southerner during her travels. Nor did she write openly about the South. Nevertheless, her region of origins can be descried in her book. In respect to content, it is evident in the culinary practices to which it alludes, as when she collects a "great pile of cauliflouer greens" from an outdoor market in Paris—to be paired with pork—watched by French shoppers "with hands on their hips trying to study what I could do with those big leaves" (17). More fundamentally, the South is evident in the book's language. As noted, Harrison's vernacular has elements recognizable as the speech of an African American who came of age in the South, and the South can also appear in her figurative expressions. She rather startlingly remarks of a white American woman that she treated her daughter "just like a black slave" (28).

The institution of slavery also shadows, in a more complex way, her portrait of the mistresses of a freezing cold household in Seville, harsh taskmasters who deny their maids adequate bedding and insist they scrub the floors on their knees. "The many fat Ladies with so many cloths on made them look like bales of cotton setting in the chairs" (201), Harrison observes. Her arresting metaphor—which recollects the Southern grotesque trope of giant women that Yaeger has traced across Southern women's writing—figures upper-class white women as outsized, immobile commodities even as they control the physical movement of poorer women, with the imagery of cotton bales both connoting extreme whiteness and hearkening back to slave economies.[20] In her text, Harrison recruits the iconic crop of American slave plantations to indict idle Spaniards, and in life she refused the job they offered her, with its evocation of her Southern past. As a self-reliant expatriate, she could make choices available neither to domestic servants in the American South nor to local workers in Spain.

Her most substantive, although still less than direct, representation of the South occurs in the book's opening pages, in which she transcribes a conversation she had during her outbound voyage to England: "In the Tourist Third are a Young Student Doctor from a town call 'a Way cross Georgia' Ga. and He keep very much to himself. I ask him why and he say he do not care to mix with emigrants. I said these are respectful bisness people going home to visit. He had never been away from his little Georgia Town and read about emigrants at Ellis Island" (2–3). Emphasizing his origins by repeating "Georgia" three times, with gracious condescension she refers to the sole Southerner to appear in her book as a "poor kid" and "my little Georgia Doctor" (3), an ignorant provincial whom she must school on the composition of American society. Her attention to his antiemigrant prejudice displaces any discussion of anti-Black prejudice. We can wonder how he viewed his fellow Southerner, but regardless, Harrison makes clear that this intolerant young man had no power over her.

Her description of a meal she relished in Ireland includes an almost ethnographic reference to her community of origins "Now no body can cook cabbage to beat the Irish of Cork not even the American Colored Southerners" (8). While she refers in passing to her African "blood," if she still viewed herself as an "American Colored Southerner," she chose not to say so. In identifying the South as the nation's "representative black space," Emily Lutenski supplies a possible explanation for Harrison's suppression of the South, as equivalent to a reluctance to discuss American racial politics.[21] When writing her white correspondents, she avoided mentioning race in a US context, which means she avoided the South.

Echoing other African Americans in France, Harrison appreciatively remarked of Nice that in searching for accommodation, "Here you never think of your color" (21). The shadow of this endorsement is the implication that elsewhere she often was obliged to "think of" her race. However, she does not document her encounters with racial prejudice in any of these unnamed places. Instead, her narrative tenders a running catalog of the many instances in which her nationality or ethnicity was misconstrued. Outside the United States, Harrison was variously deduced to be Chinese, Japanese, Arab, Cuban, Moroccan, Indian, Jewish, French, Spanish, Argentinean, and Greek. The question posed to a mixed-race character in Jean Toomer's *Cane* resembles those she routinely fielded: "What is he, a Spaniard, an Indian, an Italian, a Mexican, a Hindu, or a Japanese?"[22] Yet unlike *Cane*, Harrison presents the interpretive difficulty she embodied as having wholly positive outcomes. Her book does not represent her Blackness as a condition that is so faintly visible as to fortuitously enable her to escape it. Rather, it formulates her Blackness as a felicitous state of receptivity that promotes the assumption of multiple, cumulative identities. "Harrison wasn't passing," Korey Garibaldi insists. "She was refusing to be simplified."[23]

Her flexibility yielded both material and subjective benefit. She once stated, "I am willing to be what ever I can get the best treatments at being" (75), and numerous scenes in the book show her to be not only "willing" but also proactive, expediently adjusting her physical and social appearance through clothing, language, and behavior. On arriving in Boulogne, for example, she alleges, "I looked so much like one of the Fishman wives even the coustom offices refused to look through my baggage. most of the women have long hair and dress it in two brads as I do all I laked was ear rings I had 2 pairs in my case Mme. gave me. Well I put on my correll ear rings and was a perfect Boulognenesser" (14). As in this instance, invariably she formulates the practice in cultural terms. She passes as local, an insider. Such posturing of course has an implicit racial dimension: To be "a perfect Boulognenesser" suggests a white identity. Yet the habit was not restricted to Europe, and indeed was more prevalent in Asia. As demonstrated throughout Mollie Godfrey and Vershawn Ashanti Young's edited volume, *Neo-Passing: Performing Identity After Jim Crow*, "passing" can take innumerable nonracial forms. Godfrey and Young give examples that include "a gay person who presents him- or herself as straight, a poor person who presents him- or herself as rich, a white author who uses an Asian name." Such "identity performances," moreover, are often unstable and ephemeral, with the subject "slipping from pretending to passing to identifying and back again."[24] Continually shifting through different physical and cultural environments allowed Harrison to continually recast her social presentation, even as doing so foreclosed enduring affiliations with larger, identity-based communities.

Her publishers took her claims at face value—or pretended to. According to Morris, she was "accepted as a native" (xi) in the countries she visited. This is just ridiculous. Both her actual practice and her representation of it are far more nuanced and witty than this bald assertion suggests. Harrison's statements about her local status are not to be taken literally, but rather as affirming her fluid adaptation and ready attachment. The Japanese she meets in third-class rail cars, she claims, "forget" she is American and "begain with me as though I am a Countrymen of Theirs" (286). "Its only the rich can buy butter," she comments in Seville. "To we Sevillans butter is to dear" (205). Even more playful: "I leave on the 20th Nov. for Shanghai so will become Chinese on the 21" (291). Walsh describes such incidents as "celebratory moments of escape from the stifling racial atmosphere into which she was born."[25] Rather than attempt to convince other people that she is an actual member of these groups, Harrison makes gestures of affiliation through wearing their dress, adopting their ways, or simply seeking out their company. Assuming such guises supports her claim to world citizenship. Encompassing and transcending all her identities—whether produced by race, location, occupation, gender, nationality, or class—she is a "true rover" (126).

All that said, however, her text cannot entirely repress her Southern past and racial history and the subjectivity that they engendered. Her representation of the two train accidents she survived is once again instructive. Readers are placed deepest in the South in the context of disaster. As discussed in chapter 1, the collision outside Brno triggered memories of the Alabama derailment a quarter century earlier. She recalls, "I was in a Reck in 1903 on Sept. 1st when I was a little girl also a Monday but it was twice as bad" (53). The accident in the United States was a Black tragedy, both in respect to the actual victims and the way it was racialized in public discourse—such as the sensationalist report, "The dead bodies of the Negroes were scattered in every direction." Conversely, the accident in Europe was a white one, both in respect to the actual victims and the way Harrison chose to frame it. "All the phesants came running across the fields. they looked as clean as pins with white handkerchiefs on their heads," she recounts. "It was a hot day and it happen in a turnip patch we used the turnip tops to put under the engured ones heads" (51). The tale grows harrowing: "I stayed with a German Girl that had been torned into thread from the waist down She lived a half hour in that half hour I just loved her she were not more than 21 and had beautiful great blue eyes. she were well dressed and had a beautiful engagement ring on" (51). This is a rare moment in the text in which she lingers on a white woman's beauty. She highlights the German's identity—so unlike her own—as rich, fair, and betrothed to drive home the injustice of her death.

The premise is underscored by her startling closing words, which hint that trauma reanimates core beliefs: "Finally she died on my arm. had I been killed it would have been absolutely nothing compared to that girl" (51). Whether or not the statement reflects an internalized sense of low caste, at the least it expresses a conviction that an upper-class, beautiful blue-eyed woman should not suffer so. *My Great, Wide, Beautiful World* is both a transnational book and a regional one. In contrast to its many direct references to past events in the East, West, and Midwest, the South is only implicitly represented in the text, but it is also more deeply registered there, as a site of prejudice, exploitation, and death.

Harrison controlled her subjective and social placement through traveling. With even more lasting effect, she did so through writing, too. In the context of African American travel writing, Tim Youngs argues that "the writing, like the travel itself, is a process through which the self moves"; "travel and its textual representation combine to create a space in which identity can be affirmed, discovered, or renegotiated."[26] Harrison's actual travels and her travel narrative were reciprocal and had similar effects.

In her day many reviewers characterized Harrison as "illiterate." The charge of illiteracy is an illogical one to make of the author of the very book one reviews, but the concept was used loosely, to refer to her nonstandard English and departure from formal conventions of authorship. If one "spells by ear and knows no rules of grammar" (x), as Harrison's editor maintained of her, then one is not fully literate. Yet throughout her travels she was immersed in print culture. We see her buying books, haunting libraries, poring over guidebooks, placing newspaper advertisements, scanning job announcements, and filing references. Collectively such acts were the keystone of her world enterprise. Catherine Hobbs argues that literacy "denotes not only the technical skills of reading and writing but the technical—or rhetorical—knowledge of how to employ those skills in the context of one or more communities . . . a level of literacy that enables the user to act to effect change."[27] Outcomes include "economic gain, social status, freedom."[28] Assessed in this manner, Harrison's literacy can be conceived as not marginal but momentous.

Harrison was "studing the book that means everything to me"—her copy of *Bradshaw's Continental Railway Guide*—just as her train crashed in Czechoslovakia. Her shocked reaction shows her attachment to the volume that helped her access the experiences she craved: "I thought about nothing but this book and kept calling my book, my book I was stunted" (50–51). The textual acts she subsequently effected prove her adroit literacy. Having received medical care in Budapest and spent time recuperating in Belgrade, her final step was getting compensation from the railroad for the head injury and black eye she suffered, something she was surely not in a position to press for as a girl after the comparable accident in Alabama. She recounts, "I had a doctor to write up my inguries. Glad I found one that has Jewish blood because they know what to say the doctor told me You are traveling in that country for plesure they have no right to give you such a black eye as this. they must pay you something" (52–53). This is one of the text's two jarring antisemitic moments; soon afterward, spending several weeks in Jerusalem and making Jewish friends in Egypt seems to have moved her past this casual prejudice, or at least from voicing it. In addressing the envelope in which she sent her request, she neglected to "put all the little ups and downs over the letters" (116), the diacritical marks, and it was returned. On resubmitting her bid she asked for more money, $200, which to her astonishment she received. "Never have I got something for nothing before," she stated, continuing, "and I dont feel I can use it for Globetrotten" (116). (I think she did though.) Moving among several languages, over the course of this process she got the doctor's statement translated, wrote a petition in English, and learned to correctly address the envelope in French. The incident reflects not only her struggle with orthographic conventions but also her adaptability and growth.

As we know, Harrison's expedient personal relationships hinged on effective writing. Of a fervent scene at a train station in Turkey, she recalled, "There were women groaning moaning shaking theirselves all sitting on the floor I ask what was the matter It was because one of the Family was going to another town It was like going to America to them just a nights journey but they might never see that one again they dont read or write so its just like a peson was dying no way of getting in touch with them" (65). Her own experience was just the opposite. For these women, a loved one's "nights journey" was "like going to America." Harrison had actually traveled from America, but through diligent personal correspondence she sustained and developed her social ties.

This correspondence supplied not only steadfast emotional support but also occasional material support, the gifts and smalls sums that arrived with her mail. Harrison openly assessed whether her letters merited their expense. Sometimes she decided negatively, confessing from Shanghai, for example, "When I think of the good things I can get to eat for what the stamp cost I just stick these letters in my case" (293). At other times, however, she deemed her correspondence crucial enough to make sacrifices. She commented in Germany, "I had to send so many cards to many people about the world that have been kind to me which is a small bit to do. but it have amounted to dollars with stamps I wanted to stop at Heidelberg but cut it out so to have it for postage" (262). Her sense of epistolary obligation to "many people about the world" indicates the organic growth of her correspondence. It also shows that she conceived of it as recompense, more often a gift to repay past kindnesses than a bid angling for more. Friends she made in India and Burma asked that she write to them, as did the Israeli woman who took her in for a night and the Swiss men she met while passing a night sleeping outside.

Most of course did not become long-term pen pals. Harrison was exacting about the letters she read, once commenting of a young admirer, "He is so careful that he may always write just the pleasing things I like" (FL3). Those who relayed a deficit of "pleasing things," we can imagine, soon ceased to hear from her.

Her report on the cache of letters that awaited her at *poste restante* in Marseille shows how assiduously she attended to her mail:

> I took the tram down to my same little home of a year ago.... Then to the Express and received nine beautiful letters. When it is a long time since you have received a letter you say a little prayer in Your heart that the letters will be of happiness only.... So of the nine each had beautiful news. Two from Mrs. M. I worked for in Paris one forward

from Bombay. She always enclose a new dollar bill I always kiss bills the minute I draw them out. (166)

She exhibits her usual blend of affection and pragmatism, heartened by "beautiful news" from "Mrs. M." (Mary Morris) even as she kisses the money she sent her. We can also see that the exchange worked both ways. Far from an act of noblesse oblige, the correspondence was evidently meaningful to Morris, too, who sent out regular dispatches to *poste restante* locations without having them read, much less answered, for weeks on end. The revelation invites us to consider the role Harrison played in the lives of other, more affluent women, one that ranged beyond that of a competent servant or even a worthy cause.

Above all, the many social and material benefits aside, rendering her experience in print was a source of personal fulfillment. Conducting a capacious correspondence afforded her a method or even a pretext to write. We know, moreover, that Harrison sometimes wrote about her experiences outside the frame of correspondence. In an easily overlooked disclosure, she mentions showing Mildred Morris "many little bits that I had written down some of thing and many about diffent Admires that I had meet in different countrys their Empresion and such things that came to me" (FL2). Its frivolous subject matter—her suitor bounty—may distract us from the statement's key revelation, that Harrison's writing could take the form of synthesis as opposed to just real-time record. Having Morris review this work in progress points to her authorial pride and even sense of vocation, along with the role the other woman played in fostering it. Indeed, it was this catalog that prompted Morris to deem her "a born writer." Harrison then added, displaying her typical resistance, "Of course She was much pleased with me but I did not let it get into my life planes."

This chapter opened by posing the twinned questions of why Harrison traveled and why she wrote. I hope I have indicated the cluster of economic, social, and psychological motives that drove her prolific output. Through writing, she could conduct personal relationships, collect small alms, manage employment and travel, and craft forms of private and public selfhood. Making connections across far-flung locations, she organized her life experiences into a semblance of plot, creating meaningful narrative out of the events that accrued during her perpetual quest for freedom. Her urge to travel and urge to write were one and the same.

CHAPTER 5

Work and Play in France

In 1896, nearly forty years before Harrison's debut in the same magazine, the *Atlantic Monthly* published Booker T. Washington's "The Awakening of the Negro." The thrust of the essay is that African Americans should focus on attaining not a classical education but a vocational one. Washington opens with an anecdote about the impression made on him as a child, shortly after emancipation, by the sight of "a young colored man, who had spent several years in school, sitting in a common cabin in the South, studying a French grammar." The man's "knowledge of French and other academic subjects notwithstanding," his home troubled the future educator for "the poverty, the untidiness, the want of system and thrift, that existed about [it]." Studying French struck Washington as the ultimate intellectual frivolity. While conceding that Black men and women had "a right to study a French grammar," he maintained that "in the present condition of the negro race in this country there is need of something more."[1]

Yet even as Washington posited the pursuit of French as the wrong kind of uplift—at least for poor rural Southerners—that's exactly the form that Harrison chose, long before she went to Europe. In the preface to her book, Mildred Morris identifies instruction in languages as her most salient formal education, stating with a shade of Washington's incredulity that she "ambitiously took up the study of Spanish and French" (x). Within the book itself, Harrison reveals she had a full year of language classes in Cuba and another season of them in California. She remarks, in her first entry from France, "I am certonly enjoying the labor and money I put in French lessons the year I took in Cuba and the winter in Los Angeles" (15). In reference to daily activities such as marketing, she continues, "All that help me to speak French" (15). By the time she left the country in 1934, by all signs she was fluent.

Her studying the language in Cuba, circa 1917, proves her early resolve to go to France. This dream destination eventually became one of her home bases, joining New York, California, Hawaii, and Argentina as a place where she lived for years and to which she periodically returned. On arriving in England in July 1927, she quickly found an agreeable housekeeping job in London. However, in November of that year, fed up with the English climate, she moved on to Paris. Paris was a reliable place to get work, and the notices she placed advertising her services prompted flurries of replies. Indicating the city's overlapping practical and emotional functions, she remarked, "When I am roveing I feel sad when I think I must get back to Paris and on a job then when I arrive I feel sad for staying away so long" (243).

Cumulatively, she lived in France for somewhat more than four years. The French Riviera, including Nice—the regional capital—and Antibes and their surrounding villages eventually displaced Paris as her base. She briefly worked on the coast before departing for further travel in the Middle East and India, returning to Paris in the spring of 1930. The following winter, she was again in Paris—editing her book manuscript with Morris—but thereafter settled on the Cap d'Antibes.

France played a special role in Harrison's life, as a refuge and resting place, as a site of creative achievement and remunerative work, as a pleasure ground, and as a cultural field where she negotiated her identity as an American and as a Black woman. Her approval was boundless. In France, she declared, "Always there are delightful things to enjoy both for the Christian and sinner" (FL2). She assessed her first Riviera interval as "the most gelouries of all my life just the same plesure as when I am traveling" (26). Later, expressing her nostalgia for Gaul, she exhibited none of her usual cavalier attitude about the easy replacement of one place for another. Indeed, one wonders whether she ever would have left, had not the threat of war compelled her to.

What experiences and exposure contributed to her early interest in this country? Possibly it was sparked in Havana, so close to the former French colony of Haiti. Her proximity to French culture continued during her early 1920s residence in Louisiana, most likely New Orleans, with its Creole legacies. As Emily Lutenski summarizes, "French, Spanish, and U.S. rule; Caribbean immigration, including from Haiti; indigenous populations; and African chattel slavery all shaped Louisiana. As a result, it developed a complicated, multiracial social structure that stood in stark contrast to the racial bifurcation of 'black' and 'white' that was common elsewhere in the United States."[2] Regrettably, Harrison left no comment about her views and experiences in this unique society. Its racial variety rivaled the gradations of color that would soon delight her in India and Burma, where she "had much joy noticing the difference" among the women she saw, "some fair some light and some dark" (154, 150).

One can posit catalyzing French influences in other American locations, too. In Los Angeles Harrison became friends with Alice Foster, born in Mississippi to a French mother, and the older woman's heritage may have fanned her interest. During her employment with the Dickinsons, the real estate moguls traveled widely in France, and hearing about their trips could have added new sites to her geography of aspiration. Paris loomed large, moreover, in Black internationalist practice and imagination, especially in New York, so her time in that city may have played a role. New York was home to prominent African American writers, including Gwendolyn Bennett, Jessie Redmon Fauset, Langston Hughes, Claude McKay, and others, who traveled to Paris and wrote about their experiences. Summarizing Raymond Williams's contention, Brent Hayes Edwards explains that "the European metropole after the war provided a special sort of vibrant, cosmopolitan space for interaction. . . . boundary crossing, conversations, and collaborations that were available nowhere else."[3] Although Harrison had no known contact with these luminaries, their reports could have filtered back to her. Certainly in New York she heard about the warm welcome African American soldiers enjoyed in France, perhaps even from the army man with whom she had a long relationship.

Yet the source of her longing may well have been just the nation's glamorous image in the United States, especially that of its capital. By the early nineteenth century, Joan DeJean explains, "Paris had won out over all its obvious contemporary rivals—Venice, London, Amsterdam—and had become universally recognized as *the* place to find elegance, glamour, and even romance."[4] World War I, moreover, triggered a rush of pro-France sentiment in the United States. France was utterly unlike small-town Mississippi—hence Washington's conviction that speaking its language was so incongruous for a young Black man in the South.

During Harrison's last job in England, her elegant Italian employer, the "Countess," proposed she accompany her to Paris to assist her at a wedding. Her condition was that she promise to return to London afterward. Harrison was unwilling to travel on those terms, and instead she went to Paris on her own. Threading the needle of maintaining ties with her employer even as she resisted her control, she traveled on the very same day as the Italian, but alone and on a different route. She did indeed help her at the wedding, but she did not go back to England. They kept in touch—nearly two and a half years later Harrison noted in Madrid that she had "rec'd a Letter from my Lady of London" (213).

On landing at Boulogne-Sur-Mer, she wasted no time in miming a local identity: With her "two brads" and "correll ear rings," she alleged, she was "a perfect Boulognenesser" (14). As discussed in chapter 4, by these gestures of affiliation she could both take the edge off her difference in her lived experience and in her text make a case for her world citizenship. From Boulogne she swiftly proceeded to Paris. Her first employers were the widowed Mary Morris, "a very nice Lady of New York" (16), and her "Two Grown Daughters," Mildred and Felice. Harrison happily reported, "I was looking for a place with lots of Francs lots of time off and little work and I have that." Little did she know that ultimately her compensation—a new identity as an author—would prove far more rewarding than bountiful francs and beaucoup free time.

Retired actors all three, the Morrises were thoroughly international. Mary Morris had been married to Felix Morris, a British vaudevillian and character actor. Disowned by his upper-class family, Felix established his career in London and Paris before moving on to New York, a "busy if unheralded comic."[5] Prior to his death from pneumonia in 1900, Mary—variously identified by her married name, Mrs. Felix Morris, and her stage name, Florence Wood—collaborated with him on musical arrangements and acted in his productions. She also appeared onstage with Mildred, who had juvenile roles in London and Broadway shows, including *The Little Princess*, *Richard III*, and *Peter Pan*. Mildred played Wendy in the latter, sharing equal billing with Maude Adams, J. M. Barrie's original Peter Pan and rising star. While living in Manhattan in 1920, the trio was affluent enough to employ two live-in English servants.[6]

The Morrises were essential to Harrison's authorship. It was they who first bruited the idea of a book, and later they helped make that idea a reality. Throughout her travels Harrison wrote to Mary, and Mildred went on to serve as her editor and agent. Harrison usually referred to the latter as "the Writer." In her preface to *My Great, Wide, Beautiful World*, Mildred explains, "Her first position in France was with my mother, Mrs. Felix Morris, who suggested to Juanita that her written experiences might interest a larger public than her immediate friends. It was then decided that I should arrange her Odyssey for publication when she had travelled more extensively" (xi). Felix's early vaudeville career, which saw him praised for "his talent in refined dialect parts," may have helped the Morrises recognize the commercial potential of Harrison's vernacular prose.[7] Crucially, Mildred had worked for four years in New York as staff correspondent for William Randolph Hearst's International News

Service, specializing in women's issues. Following the family's early 1920s move to Paris, she continued to write freelance articles, covering high-brow cultural events such as opera. Her freelance acumen enabled her to pinpoint the most arresting content of Harrison's letters and to recognize the *Atlantic Monthly* as a potential venue. She may even have had a personal connection to the magazine.

That project, however, was still in the future. During Harrison's first months in Paris, the city's plenty absorbed all of her attention. With the passing of postwar austerity, Paris had reemerged as an exuberant center of art, literature, music, dance, and theater. The revival continued unabated until the global depression reached France, somewhat later than the rest of Western Europe. Due to the heady mix of cultural ferment and economic boom, the 1920s in Paris have been dubbed *Les années folles*, "the crazy years."

Inseparable from this rejuvenation, the city became a center of pan-African community and activity. Petrine Archer explains, "Paris became the European hub for blacks from all over the diaspora, including Africa, the Antilles and America, where they discovered their commonality with each other in ways that had not been possible in the United States. Guided by thinkers such as W.E.B. Du Bois and Marcus Garvey, this intelligentsia used the city as a base for the promotion of a pan-African unity."[8] In 1919, Paris hosted the first Pan-African Congress. It was also home to the "Negritude" movement, which sought to foster solidarity among French-speaking Black intellectuals. Such views were not confined to a cultured minority: Archer states that inhabitants of the city's working-class districts, too, "developed their own sense of being part of a wider diaspora that was creative and cosmopolitan."[9] Harrison was among them, putting into lived practice the international outlook that theorists espoused.

James Weldon Johnson's reminiscence is a potent early expression of what made Paris feel special to African Americans: "I recaptured for the first time since childhood the sense of being just a human being.... I was suddenly free."[10] Despite his grousing about Parisians—too drab, too stingy, too ill-natured— Hughes allowed, "But the colored people here are fine. There are lots of us."[11] While fewer in number, African American women such as Bennett and Fauset responded positively, too. Fauset wrote in a 1925 essay in *The Crisis* that in Paris she could have tea "*at the first tearoom which takes my fancy*" (original italics).[12] "It is lovely just being oneself and not bothering about color or prejudice," she remarked in a letter to Hughes.[13]

Fauset also once stated, "I should like to see the West Indies, South America and Tunis and live a long time on the French Riviera."[14] Tunisia aside, Harrison fulfilled all of Fauset's geographic dreams. Despite their quite different class locations, these two writers shared considerable common ground. I have yet, however, to find statements about Harrison from Fauset or other of her African American women contemporaries (save Era Bell Thompson), although they surely exist. Ironically, given the highly gendered nature of Harrison's experience as a "vagabond housemaid," the most accessible commentary is from men.

While unlikely to have had direct contact with Harlem notables such as Hughes and Fauset, Harrison certainly enjoyed Paris's nightlife—reporting with some astonishment on the scantily clad dancers at Moulin Rouge—and an entertainment scene that was energized by African American performers and entrepreneurs. In the 1920s France was gripped by "Negrophilia," an avidity for Black cultural expression, especially music and dance. A character in Fauset's *Comedy: American Style* muses on "the traditional fondness of the French for the Negro."[15] The trend, rather incongruously, was a by-product of avant-garde enthusiasm for African sculpture, the spoils of colonialism.[16] The African American veterans who remained in Europe after the war, especially those who worked as performers or opened clubs of their own, further whetted Gallic appetite for Black culture. These expatriates were joined by Black musicians and dancers on the club circuit. The most famous was Josephine Baker, who starred in La Revue Nègre in 1925; Harrison sent Foster a notice she had clipped about Baker's pet panther. Jazz singer Ada Smith relocated from Harlem to Montmartre, where she opened the legendary Chez Bricktop.

Historian T. Denean Sharpley-Whiting states that aside from the touring performers, "there were very few African American women, let alone artists and writers, in Paris in 1925."[17] Although the two newspaper clippings she sent Foster from Paris prove the special interest she took in African American women visiting that city, Harrison's book reflects this relative scarcity. The majority of the "colored" people, to use her adjective, that she notes seeing in France were men. Summarizing her charting of its arrondissements, this American *flaneuse* reports, "I have been in every corner of the city and all by foot" (15). However, other Black women out and about on the French city streets do not feature in *My Great, Wide, Beautiful World*.

By February 1928, she was ready to move on. "I told Mrs. Morris I felt I must go now," she disclosed, adding, "She want me to come back. I am going to try to" (19–20). Her immediate goal was to reach Nice in time for Mardi Gras, but India was in her long-range sites. So too was the prospect of authorship. Discussing her possible return to the Morris household, Harrison confided, "One of the Daughters is a writer and the mother said my travellers should be put into a Book. I told her I would come back after my trip to India and work for nothing if Miss Mildred, the Daughter would help me" (16). It is striking that the Morrises broached the subject scarcely half a year after her first arrival in Europe. Their early interest suggests that not only her recent continental experiences but also her previous ones in North America made them recognize her life as worthy of a book. Of note too is Harrison negotiating for the editorial work to make it happen, offering unpaid domestic labor in exchange.

She proceeded to travel for five months in Europe and the Middle East before reaching India, followed by nine months living in Spain, experiences discussed in later chapters. English had yet to displace French as the language of diplomacy. "My French serve well" (48), she remarked in respect to paperwork at consulates across Europe, and it became even more serviceable when she visited former French colonies.

When at last back in Paris, she observed, "Paris seem much quiter than two years ago or it may be myself after going around to the gayest places for such a long time it seem quite" (243). During her reunion with Mrs. Morris, the older woman repeated her suggestion: "She said my travellers should be put into a Book. just as I have written them misteakes and all" (243). Yet its time had still not quite come. Rather than rejoin the Morrises, Harrison briefly took a job with a Russian American who had sought to hire her in the past. The woman's eighteen-room apartment was located on the Avenue de Fochs, the grand boulevard leading into the Arc de Triomphe. It was crammed with priceless art and antiques, and yet the sugar, cheese, and napkins were locked up. Of her difficult employer, Harrison relays, "One thing she like as She have travelled is to hear of my travelling and think its wonderful when I am so poor. but she often gets my got. so I say things to her it seems to help" (246). Placating a woman in power with picaresque tales was her forte.

Next she entered the household of Captain and Mrs. "C.," as she refers to them, an American military couple. The job opened up because Mrs. C. had disliked her local maid. "French maids are not as good as they are wrote up to be" (249), Harrison wryly noted from the vantage point of an American one. The captain, a member of the Army's Quartermaster Corps, was among the staff that ran the ambitious Gold Star Pilgrimage program, organizing trips to Europe for mothers and widows to visit the graves of fallen relatives. Close to sixty-seven hundred American women were brought to France, all expenses paid by the US government, with Paris the base for their tours.[18] The first group, "Party A," arrived on May 6, 1930, just before Harrison was installed at the C.s'. She once commented that her past relationship with the army man made her feel "at Home" (102) at American military commemorations, and it likewise would have attuned her to the C.s' work.

Yet for all its well-funded generosity, the program was racially segregated. The 279 African American participants were organized into six all-Black cohorts.[19] Separate and unequal, prior to embarkation they were lodged at the Harlem YWCA while their white counterparts stayed at the imposing twenty-two-story Hotel Pennsylvania in Manhattan; they crossed the Atlantic on the SS *American Merchant*, which doubled as a freight ship, whereas the others took bespoke passenger liners.[20] Rebecca Jo Plant and Frances M. Clarke state that this federally mandated discrimination "ranked high among the concerns that preoccupied black journalists and activists in 1930s," who deemed it "the crowning insult."[21] The NAACP urged a boycott. Twenty-three of the African American invitees refused to go, and countless others agonized over the decision.[22] John W. Graham notes that the media attention accorded the Black visitors was also problematic, with reports of their visits to memorial sites displaced by "enhanced coverage of food and parties."[23]

Harrison does not allude to the controversy in giving her impressions of the first group of "colored mothers" (251) to reach Paris, only observing, approvingly, "the 55 colored were all nice fat Mamas one thin" (253). The C.'s had made a point to aid her in greeting them, giving her money for a taxi to the station. In accord with Graham's contention that the visitors were "embraced" by "the sizable African community in Paris,"[24] she reports that "the Leader of a noted colored jazz ban that are engaged at one of Paris leading high class Resturant asked if they could welcome the mothers with their band" (251). The event was held at the Hotel Imperator, where "Party Q" was lodged. It was organized by Ada Smith and Noble Sissle, whose orchestra greeted the women with a rendition of Al Jolson's "My Mammy."[25] (Defending Jolson, who had a predilection for blackface, Sissle maintained that he "immortalized the Negro motherhood of America as no individual could."[26]) One of the clippings that Harrison sent Foster describes the visitors as "pleased 'Mammies'" who "like happy children" "shrieked with laughter."[27] Any comment she may have made about this depiction has been lost.

Following a three-month stint of work in Paris, she hit the road again. She took the waters at Baden-Baden, visited Cologne, toured Verdun, and saw the Oberammergau Passion Play. Like the Morrises before them, the C.'s had hoped that she would return to them after her holiday. She writes, "I am proud to have Kind People to Want me to be with them so I made a promise to come back in two weeks like the promise I made to Mrs. M. but I dont think it will be two years. But qu'n sabe (Spanish)" (252). Yet if she did return, we have no record of it, since after that trip her travelogue abruptly ceases.

My Great, Wide, Beautiful World includes no entries between late September 1930 and the summer of 1934. A terse editorial statement accounts for the gap: "Because of the unsettled condition in China Juanita determined not to complete her journey around the world at this time but returned to the south of France where she remained until May 1934" (265). China was in the midst of a protracted civil war between the Communist and the Nationalist Parties, and Manchuria was occupied by Japan. Her publishers were correct that touring China at this juncture would have been ill-advised. However, by then Harrison had lost her investment income, the likely reason why she stopped traveling for several years. As George Dickinson, heedless of her privacy, informed the *Atlantic Monthly*, "After the crash in 1929 the trouble commenced. . . . Now, instead of having her money invested in what we supposed were good mortgages, drawing good interest, she has but little interest." She also had to cover expenses for her "vacant lots."[28] The reversal stemmed from more than just bad luck. It also stemmed from Harrison's tendency to gravitate to economic hot spots, for the most part by physically moving to them, but in this case through investing in a real estate bubble.

Showing how much she had relied on the income, as she prepared to leave India in July 1929 she divulged, "I must get back to Europe as my checks are getting very low" (146). The following week, she asked Myra Dickinson to forward her "a little interest money" so that she would have something to live on once she reached Marseille. Should there be none, she requested that she "kindly stick in a love offering," promising to "keep an account and return it" (150). It took some time before she found out about the new state of her finances. As late as December of that year, she happily wrote from Seville that the American dollar was "high," albeit not "high enough" to "keep" her (200). Soon enough, however, she would learn she had no more dollars coming in, her response to the unpleasant discovery not on record.

The long delay between the Morrises' proposals that she write a book—in December 1927 and then again in May 1930—and their starting the project proved fruitful. Harrison's ongoing separation from friends and mentors drove her voluble correspondence, while the experiences she had in the course of the separation supplied the content. The result was a solid corpus to draw from in composing the manuscript.

Finally in January 1931 it was time to tackle the book. To that end she moved into a commodious apartment on the floor just below the Morrises at 39 Rue de la Bûcherie, lent by their accommodating out-of-town neighbor. Again showing Harrison's habitual occupation of prime real estate, the "beautiful room" in which she and Morris worked gave onto a "restful view of the Seine and the Cathédrale Norte-Dame" (FL2), in her words. Nowadays there are usually groups of tourists milling about in front of it, as her former workplace is just next door to, and actually within the same large edifice as, George Whitman's Shakespeare and Company Bookstore. Whitman's establishment is named after Sylvia Beach's English bookstore, receptive home to between-the-war writers including Ernest Hemingway, F. Scott Fitzgerald, Djuna Barnes, and Gertrude Stein. In her apartment at 27 Rue de Fleurus scarcely a mile away, Stein wrote an unlikely memoir of her own, *The Autobiography of Alice B. Toklas*. It was first serialized in the *Atlantic Monthly*, and Harrison and Stein shared the same editor, stuffy Ellery Sedgwick. As I have argued elsewhere, publishing Stein's deliberately vernacular narrative prepared Sedgwick to accept Harrison's actual vernacular one.[29]

With her usual safeguarding of autonomy, Harrison turned down the Morrises' invitation to take her meals with them. "Their hours didn't suit me," she informed Foster, "as I hate food so much and eat to live not live to eat" (FL2). Was she joking? Her book is studded with scenes of joyful consumption, such as in Cairo on Thanksgiving Day, 1928: "I went to one of my many eating places and had native beans and rice tomatoes and young salad. I felt good, and of all the things in the world Ice cream only can give me an elegant feeling so I went to Groppi who make the best icecream" (82). Yet her refusal of the Morrises' offer is revealing in other ways. A ready conclusion is that she was disinclined to dine with the family she had once served, with racial and class differences making the prospect feel fraught. However, in describing the alternative she fashioned—eating out midday, with the Morrises obligingly delivering breakfast and supper to her room—she invites us to see her partiality for modern forms of sociability as the real reason. She explains, "I like being among artist so took my lunch at a lovely small home like Russian French resturant in just two blocks on the same street a good wholesome 4 course lunch for 2 dollars 40 cents 10 tickets for 60 francs. and there I had such a good time as I meet ones from different parts of the globe and which I often was no stranger" (FL2). The intimate bistro afforded a worldly, bohemian company that the Morrises at home could hardly rival, an assembly among whom she revisited in spirit the "different parts of the globe" she had come to know. Left to her own devices, she could do better than eat with three American women in private domestic space. As Morris herself relayed, when still a child in Mississippi Harrison concluded, "No one is there for me to copy, not even the rich ladies I work for. I have to cut my life out for myself" (ix). Her dining solution is one of countless examples of how she "cut out" her life.

Morris was well placed to serve as Harrison's editor. In the capacity of staff correspondent for Hearst in New York she had written sympathetic stories about subjects that include chorus girls on strike, murderers on trial, suffrage activists, the Women's World Disarmament Committee, the Bill of Women's Rights, and the fashion trend for the "bare leg." She was staunchly behind the feminists. Her 1919 article about the strikers, for example, includes this ringing quotation: "We're workers and we're ladies and we want the chance to make an honest living."[30] Morris showed little goodwill toward Catherine Eva Kaber, the "murder queen" of Cleveland convicted of killing her husband, but in writing about other homicide cases she avoided the usual sensationalism. Representing the acts and views of a host of rebellious women prepared her to do the same for Harrison. (Intriguingly, a Mildred Morris is listed as the author of "Doomed to Be Dead as Long as He Lives," the cover feature of the May 1929 issue of Bernard Macfadden's cult magazine, *True Strange Stories*. Could it be her?) Of course, the service she rendered Harrison was not wholly altruistic. Morris must have seen some professional opportunity for herself in the project, especially since her success in placing work in periodicals had tapered off since leaving New York.

By some measures, Mildred Heywood Morris and Juanita V. Harrison had much in common. Born within a decade of each other, the two Americans had both been child laborers, an actor and a housemaid, respectively. Each enjoyed a brief period of renown, and each partially supported herself through writing. Yet for all their shared mobility and single lady self-reliance, the difference between their lives is immeasurable. At age ten in Mississippi, Juanita completed the rudimentary schooling allotted her, whereas Mildred and her sister attended Hillside Home School, the progressive boarding school in Wisconsin whose building was designed by Frank Lloyd Wright for its two founders, his aunts. During the years that Harrison was developing her repertoire as a skilled domestic worker across North America—and building her portfolio of employers' references—Morris was developing her repertoire as an actor in New York and London—and building her collection of actors' autographs. Just as Morris's fame was peaking on Broadway, Harrison, in her words, began

FIGURE 3. Mildred Morris (left) with Maude Adams in *Peter Pan*, 1905. Source: Billy Rose Theatre Division, the New York Public Library. "Mildred Morris and Maude Adams in the stage production *Peter Pan*." New York Public Library Digital Collections.

living "a little like I had planned" (FL2), likely by taking short excursions from her Columbus home.

As a young woman, Morris sat for Irving Ramsey Wiles, who specialized in actors.[31] Her studio portrait, radiating high-society affect, contrasts with the impromptu sketch that artist Carl Brandien made of Harrison while visiting her in Waikiki. "As a token of their friendship," the *Honolulu Star-Bulletin* records, "Brandien painted a picture of the author's tent with Juanita sitting in front of it. In the background is the ocean and nearby are palms and flowers. This colorful picture he presented to Juanita."[32] Perhaps she posed wearing the garment described in a newspaper profile the same year, "embroidered by her own hand and looking partly like Hawaiian, partly like South Sea island dress and mostly like Juanita."[33] Or perhaps she had on the "gay shorts" she wore to her debut literary reception in Honolulu.[34] Morris, as a museum catalog details, was attired for her sitting quite differently, "in pale dress with collar and dickey painted cream with color highlights, and puffed sleeves with daubs of reds and pinks, with dark hair put up behind."[35] Telling, too, is the difference between the fates of the two works. Brandien's gift to Harrison, probably painted on cheap board, was ephemeral. Similar to the materials pertaining to her book, it was never archived and likely lost. Morris's commissioned oil on canvas portrait, in contrast, donated by Felice, forms part of the permanent collection at New York Historical Society Museum. The Morris sisters' collection of stage autographs is likewise preserved in Princeton University Library's Special Collections. Nevertheless, the unevenness of Morris's and Harrison's privilege does not erode their common ground. Independent, adaptable, and migratory, Morris, too, had to rely on her own resources as she navigated a patriarchal world.

"Her employers invariably became her friends" (x), Morris alleges, and surely she viewed as such her own family. However, she can strike a condescending note, and her preface teeters on caricature with assertions such as "she settled down on the Island of Hawaii where central heating is unnecessary and a minimum of work gains abundant food" (xii). Language such as this may indicate the limits of Morris's writing abilities; the ingrained beliefs that dictated how white people saw Black women also controlled her prose.

For her part, Harrison's record of their joint project exhibits the shifting currents of power and authority between these two enterprising Americans. She begins, "The Writer young and beautiful came down 930. a.m.," continuing,

At the begining the Writer thought that She would type a short sketch of my life for the interducting of the letters. but I knew that it couldn't be a short sketch. because my life are like the Saying to "never judge a book by its covering" some time the covering look not at all enviting but the inside are most enteresting so in an hour the Writer found it to be so much so with my life that She became so interested that she said she could not do Her other writings she was so anxious to get down to me to go on with my true life story which I begain at Six and she thought it wonderful how I could go through it as though it was the day before and it was such hard work as she did not know short hand, so had to write so fast. and often ask me how could I remember every thing so easy and where did I get the wisdom. at the end said that She had never enjoyed writing so much as the month and 10 days. I also enjoyed it Because She is a brilliant intelligent Lady. and I enjoyed watching how Her beautiful brown eyes sparkled as I went from Six years old up to Jan. 1931. (FL2)

Their relationship had changed, from that of servant and employer to that of author and editor. Carolyn Steedman has likened "telling someone else's story" to "the condition of servitude," and in assuming the role of amanuensis, it was now Morris who functioned as the attendant, not only bringing Harrison food but also waiting on her words.[36] She recorded her long oral narrative and painstakingly transcribed her letters, putting in the considerable extra work that preserving all the irregular spelling, capitalization, and punctuation entailed. That Morris "could not do Her other writings" registers the potency of Harrison's narrative. It may also indicate her growing recognition of its value, worthy of displacing her own work. Indeed, Morris is now most remembered as Harrison's editor. Even as her protégée occupies an important place in American literary history, her own publishing record can be detected only with effort.

For the project, Harrison contacted past correspondents, requesting that they return the letters she had written them—another act, absent from the official record, that demonstrates how actively she participated in the making of her book. Nonetheless, Morris was the main agent in editing and assembling them. Of the letters she retrieved, Harrison wrote, "I gave them to The Writer to look through" (FL2), continuing, "Where she had just finished arranging my letters I had more than enough to make a book." Moreover, she indicates that she worked on Morris's timeline. She figures her textual labor as tasks she was obliged to complete during set hours, describing her schedule as "930. to 12 noon then I was free for the rest of the day to do as I pleased" (FL2).

Before their work began, she had hinted at duty and even constraint in admitting to Foster that "its not easy for the writer to keep me as I am not interested although we have a contract and would like to be off for sunny Nice" (FL1). "I gave the time to the Writer all the time the French Lady were waiting for me to return," she explains, positioning the editorial work as an alternative to resuming a service job. While she had planned to be installed on the Riviera by February 1, "the beginning of the season" (FL2), she was persuaded to stay on in Paris until the 24th.

Just prior to this period with Morris, Harrison was employed by that importunate "French Lady," to whose household she promised to return once done with the book. In fact, she fully intended to hurry to Nice. She subsequently led the woman to believe that Morris had accompanied her "down here," as she put it to Foster, and that they were continuing their work together in Nice. (One shouldn't hesitate "to lie to keep a Friend" [FL2], she advised.) Far from it, she was relaxing at a small hotel near the city center, Hôtel Plaisance. Showing how her sense of class identity shifted according to environment and circumstance, she gloated in reference to this would-be employer that she was "being more of a Lady than she is altho. she is Rich and have a lovely Husband" (FL2).

The revelatory letter Harrison wrote to Foster from that nice Nice hotel ran to nearly three thousand words. She relished the establishment's steam heat, attentive service, and overall *plaisance*, explaining, "I would have stayed at the 'Y' but for the reason of this being served in bed which I dont rob myself of the pleasure for the sake of a few 50 cents. so I have a lovely cozy room in this Hotel just a few steps from the Y.W.C.A. door where I go for my meals when I want to be with the Pleasant 35 or more girls. . . . I keep buisy but I more often sleep until 10. or 11 a.m. everything are so nice and quiet in the Hotel. and so clean." She confides that as a girl in Mississippi, plotting her future, she had dreamed of doing just this: "One of the plannes in my life was when ever I really desired it to remain in bed a 1/2 day or all day and have my meals served to me by a nice maid or Waiter and in a nice way so today I am enjoying that delightful plan" (FL2). Coincidentally—or not—when I trekked to Nice on Harrison's trail, I found that my companion had booked that very hotel for us, randomly selected from a raft of budget hotels; its doors closed for good several months later. While worn at the edges, Hôtel Plaisance still had a wide marble staircase, spacious lobby, and corner rooms with balconies commanding views of the intersection of Rue de Paris and Rue Lamartine. During my stay, it was easy to imagine Harrison writing and recuperating in this tranquil place.

After those intensive forty days of work in Paris, the book project seems to have been put on hold. We do not know what interim steps were taken before the *Atlantic Monthly* excerpts and the published volume came out more than four years later. "We have a contract," Harrison stated in 1931, but did they really, and if so, of what nature? In 1934, she briefly worked for the Morrises again, giving her and Mildred an opportunity to update their plans. She then embarked on the last stage of her world tour. During those eight months on the road, the prospect of the book surely shaped her travelogue, its tone, frequency, and emphases. It also might have made her feel more pressured to produce. In Sweden, she stated with perhaps a note of anxiety, "There are no end to my Friends. I have no time to do any writing" (276).

As with her editorial work with Mildred Morris, *My Great, Wide, Beautiful World* was generated by, and inscribes, unequal but mutually beneficial relationships between Harrison and many other people, mostly women. By means of their diligent correspondence, Myra and George Dickinson structured her own, which they also carefully preserved.[37] In the dedication, Harrison credits Myra with sparking her desire to write. Mary Morris, another regular correspondent, recognized the letters' appeal to a broader readership and potential to form a book. Alice Foster not only saved Harrison's letters but also returned the majority of them to her in Paris and lent the remainder to librarian Miriam Matthews in Los Angeles. Matthews exhibited the letters along with other papers, and some seventy-five years later, Foster's granddaughter, Ann Cunningham Smith, donated them to the University of California, Los Angeles. While Harrison wrote each word, collectively these women helped create her text and secure its legacy.

Foster's contribution is the most intriguing aspect of this collaborative process, even though it is unlikely that the letters she returned made it into *My Great, Wide, Beautiful World*. In requesting them, Harrison had explained, "I have written to all the People that I wrote to and have many but cannot have to many as some may have only a page thats worth while" (FL1). Foster promptly sent off a bundle to Paris. Matthews later claimed that those same letters "make up part of the book."[38] However, they only reached Paris the day before Harrison left the city, and in any case, as she informed Foster, thanks to the Dickinsons' caretaking they already had more than enough (FL2). Some five years later, in assuring Foster that she would have Morris return "your package of letters," she repeated, "as you know they arrived after we had finished" (FL3). To all appearances, they did not enter the manuscript.

Nevertheless, that Harrison made such a request of "all the People that I wrote to" does raise the odds that the book includes letters to other personal friends and that their recipients were more various than usually conceived. In 1937, she claimed that "all the letters in my book" were written to the Dickinsons, but given that the volume includes missives addressed to Mary Morris and Helen Rose, this is evidently not the case.[39] The ambiguity about the range of original recipients complicates our interpretation of the text. With her aspirational working-class family, a woman like Alice Foster could have identified far better than an upper-class woman like Myra Dickinson with Harrison's gratification on landing a job or glee in quitting it. Readership inflects, for example, the anecdote about her abusive employer in Paris: "While You are helping her to dress She stands in frount of it looking at You in the mirrow and calling You such auful things I run out many times she call me right back" (246). Was her first reader someone Harrison once worked for, making it pointed commentary about how not to treat one's maid? Or was it perhaps a woman who herself had been in service, adding greater weight to that "You"? Either way, once the narrative became available to a public readership, it opened up to accommodate both perspectives.

During her first months in Paris, her low-key gig with the Morrises afforded Harrison the time to study German, as was her habit looking ahead to future trips. "I wanted to know something about the Paris Artists life that I had heard of" (16), she remarked just days after arriving in France, and so she made a point of seeking out a German language tutor who lived near the École des Beaux-Arts. The Latin Quarter became her favorite neighborhood. She herself liked to draw, and several of her line drawings can be seen at the edges of the book's original cover design. Her "small book of skitches," she boasted, led Morris to deem her "a gifted artist" (FL2).

"I like being among artist" (FL2) she announced from Nice. Her closest proximity with bohemian arts culture came not in Paris but in southern France, as practiced by the community of expatriates who gathered there in the 1920s and early 1930s. This population included the same "Lost Generation" writers who frequented Sylvia Beach's Shakespeare and Company bookstore, and Harrison's experience was contiguous to theirs. Complementing and countering the oft-told tales of their projects and parties, through registering the perspective of a working-class African American woman her book reminds us that there were all sorts of expatriates in France.

Harrison's Riviera geography comprised the port of Marseille, with its salient colonialist legacies and African presence; Nice, where she both held jobs and rested between them; Cap d'Antibes, the location of a superlative house-sitting gig; and tiny Juan-les-Pins, which afforded her a dependable "place" along with abundant opportunity to participate in the new vogue for sunbathing. The region's gorgeous landscapes, delicious climate, bohemian society, and, not least, bustling casinos made it a highly agreeable home.

The word "delightful" sounds a keynote in her account of her activities there:

> I have enjoyed every day since coming down I was at Cannes and there enjoyed all the delightful and beautiful entertainment at the Caisino and many delightful auto trips about and in the Alpes Maritimes.... I visited Grasse and the Perfumery as its my life to see and enjoy. Then I had a delightful stay at the Pretty little Sea side Place Juan-Les Pins. Many fine trees where it gets the name Nice I always like far more than Paris.... Nice always have much pleasure and for a few Francs. and the two big Caisino are always delightful places to spend the time from 2 P.m. to 2 a.m. or in the fore noon if one choose. but I have so many other joints to get around to and my churches to visit that I keep buisy. (FL2)

Her experience speaks to rather recent developments. The French Riviera, named as a counterpart to the Italian Riviera just across the border, had once been a health mecca favored by tuberculosis patients. By the mid-nineteenth century, newly accessible by rail, it had become a vacation destination for the hale, drawing an aristocratic crowd of Russians, English, and Germans. Winter was the high season, with visitors decamping in summer to avoid the heat and mosquitos. In the 1920s a wave of American artists, business people, and celebrities flooded the area, with the tide of upper-class European vacationers receding before it. Staying on after the high season to enjoy summer sports, they imported new habits to the coast.

The Riviera's transition from a winter to a summer resort was accelerated by the active social life of a single American couple, Gerald and Sara Murphy. Sara, a scion of the wealthy Wiborg family, and Gerald, a businessman and painter, had moved to France in search of a more liberated, creative lifestyle. They settled on the Cap d'Antibes in the roomy estate they christened "Villa America," where an illustrious group of artists, musicians, and writers gathered around them, the likes of Zelda and F. Scott Fitzgerald, Dorothy Parker, Cole Porter, John Dos Passos, Pablo Picasso, and Jean Cocteau.

Gerald recounts the act whereby they took possession of their new home: "Right out on the end of the Cap there was a tiny beach—the Garoupe—only about forty yards long and covered with a bed of seaweed that must have been four feet thick. We dug out a corner of the beach and bathed there and sat in the sun, and we decided that this was where we wanted to be."[40] Picasso's series of beach paintings include *La plage à la Garoupe*, and Fitzgerald's *Tender Is the Night* opens with a bathing scene there. And Harrison wrote about it, too. Residing "in a few minutes of Garoupe Beach" (24), as she puts it, she fashioned a miniature retreat on the cliffs above, her "favorite spot." She confides, "It is so cozy and private I call it my villa Pine needles make my bed and I bring my books to read on my coming tour. one a short Geography of the World, the other 'the Queen' newspaper Book of Travel" (29). Like her famous compatriots, she too fused leisurely outdoor life with creative enterprise and ambitious dreams. Whereas Fitzgerald dubbed the place "a playground for the world,"[41] Harrison went him a step further to deem it paradise: "Well They say that Paridice was lost but they lied because it is in this World because I have seen much much of it" (FL2).

Her account of her first sojourn there, in spring 1928, suggests that she found the region's African presence an additional attraction. Describing the journey from Paris to Rome, Frederick Douglass had reported, "As the traveler moves eastward and southward between those two great cities, he will observe an increase of black hair, black eyes, full lips, and dark complexions."[42] Writing half a century later, with like satisfaction Harrison lightly but definitively corroborates the greater darkness to be found in southern France. For example, she comments in Marseille, "There are many colord soldiers and they ware red turbons" (21). These men were probably Zouaves, North Africans who served in the light infantry regiments of the French Army, celebrated for their bravery along with their eye-catching costumes. In the same city, she spurned all but one of the many men who tried to "make love" to her, "a very nice colord one a captain or something in the Army" (21). She met the man for "5 o'clock tea" and enjoyed a quantity of French pastries at his expense, only to give him the shake by promising to meet him for breakfast the next day at a hotel she never set foot in.

She accords the closest attention to the elegant mixed-race couples she saw at the casino in Nice, reporting, "I went several times to play a few Franc in the Casino and on the Ball room floor Dark colord girls in their evening cloths dancing in the Arms of Hansom White Frenchmen" (22). She takes pains to specify the visual contrast between the dancers, "Dark" and "White," and her loose syntax leaves it open as to whether the arresting sight contributed to her repeat visits to the casino. Of an African American nurse she later met in Rome, she relays, "It seem like a dream To her to have a Hansom Italian kissing Her hand" (36). The metaphor conjures the fantastical change in social status that both women experienced in Europe, even as Harrison insists, "I hadnt give it much thought," displacing any reaction she may have had onto the nurse. As noted, in discussing her fondness for Nice, she makes her most explicit statement about her experience as a Black traveler, the distancing "you" notwithstanding: "I always get a comfortable and Home like place to stay for here you never think of your color" (21). Such tolerance was surely a factor in her liking the Mediterranean city "far more than Paris" (FL2).

Nice had been a pleasant place for Americans to economize even prior to the stock market crash. Fitzgerald's 1924 satirical essay, "How to Live on Practically Nothing a Year," describes his and Zelda's failed attempt to do just that. But for Harrison, more than just a favorable exchange rate made it a strategic choice: the large American population meant that there was a surfeit of well-paid work. Despite their reduced means, its members aspired to maintain their customary lifestyle, a bonanza for hoteliers, tradespeople, and servants. Not only locals profited from their presence. So too did the migrant workers who flocked there, such as the Englishwoman who proposed an exorbitant sum to work for the Fitzgeralds as a governess. Far less attention has been accorded this manner of expatriate than the people they worked for, filling a service gap by providing high-end services.

Distancing herself from her nationality, Harrison adopted the native outlook in viewing the resident Americans mainly as a source of income, stating, "The French have not time for them only to make them pay well for everything and I agree with the French" (24). While her portfolio, too, had taken a hit, with characteristic resilience she redressed her loss of passive income with increased wage labor. By moving to the area, she made money from an impecunious but still leisured class, while enjoying some leisure herself. Centering the worker's perspective, Harrison helps us recognize not only the differences between migrant workers and their employers but also the experiences they shared.

Her book marks her encounters with a bohemian crowd as well as members of a more staid upper class. She had several employers in the South, including a hugely wealthy American woman with not only a Riviera residence but also, according to Harrison, a "show place in Rhode Island and a winter home in

Santa Barbara Calif" (28). She admitted to "making a good number of solod francs packing them away" (29) in her employ.

With greater consequence, she also worked for Helen and Hugh Rose, an English couple who divided their time between properties in Cornwall, Mallorca, and Cap d'Antibes. Harrison first had a position at their "Villa de Verveines" in 1928, and she returned when her travels paused. She gives an intimate view of their appealing home:

> My room window looks out on a lovely Pin grove next to the Villa the Orange trees are about 30 years old and the plum trees and persemmons are lovely loaded with blossoms. The Villa were built by French Pheasants some 65 or 70 years ago and so solid and homelike we have electricity to heat also small grates. I have never felt so homelike in a house as I have in this one I sleep with my head to the North. (24)

In the midst of her extended Riviera sojourn, she reported, "I still have the joy of making this dear little Villa my home that belong to the nice English couple. They have taken a Villa for 2 years at Mallorca and they say when this Villa is Vacent it is my home. I havent had a house all by myself and when I can have it I do not mean to let it stand empty" (265–66). Thanks to the arrangement, she had worked just four months in 1933, leading her to conclude, "The last year have been the happest of my whole life" (266). That cutting back on work was the source of peak happiness suggests the toll that decades of labor had taken, while the rather plaintive comment about never having had "a house all by myself" speaks to a lifetime passed in other people' homes and rented rooms, as well as, perhaps, the cumulative exhaustion of habitual itinerancy. As she had wistfully stated on her return to Seville four years prior, "It was just like coming home not that I have ever had a home to return to but it must be something like it" (221).

The Roses' involvement in her life, while scarcely visible in the historical record, must have been considerable. The book includes a letter Harrison wrote to Helen Rose from Japan that opens "I hope that you are enjoying life as much as I am for then I know that you are happy" (291) and concludes with an appreciative nod to "the warm clock coat and fur you gave me" (292). What appears to be an actual letter of reference from Rose ran in the same *Atlantic Monthly* issue as one of Harrison's two excerpts. While its inclusion reflects the magazine's aggravating inclination to foreground her identity as servant rather than author, the letter does indicate the kind of relationships Harrison formed with her employers, as well as her personal force. "My husband and I are only too willing to give you references for Juanita," Rose wrote. "We were very fond of her and found her an amazing character, besides being a very good servant. We let her stay at the villa, knowing full well that she would look after it just the same as if we were there. Juanita is a really good woman in the finest sense of the word, and full of a wonderful humanity."[43]

Although the Roses do not seem to have been among them, Harrison had regular contact with a bohemian assembly of artists and partiers. She describes the scenes she observed with understated irony. Of an atelier tour, for example, she recounts,

> I went into one of the Studios at Old Antibes on the top floor of an old house. There was an everlasting hansom American all brown from the sea and sun there and four or five real painters sittings about one very fair in a pair of brown pyjmas and orange sweter another with slick black hair in black pyjmas and red sweter I paid no attention to the oil and paint on the walls. I expect he paid some poor but real artists to do the canvas work. (27)

She cryptically concludes, "There are plenty of scandlous durings going on here." "Scandlous" might refer to passing off other people's art as one's own. However, with its scenario of bronzed men sitting around together in their "pyjmas," the anecdote may hint at forms of gay community. She describes an older American woman and her young dreamboat French husband with a like measure of censorious fascination: "She is very rich she have a young girls shape nice silver bob gray hair dresses nice I spend much time looking at them.... She stands on their poch and watches for him if he be a little late" (267). The harshest language in the book is reserved for her "vile" American employer in Juan-les-Pins. The woman mistreats her daughter and is besotted with a dissolute poseur, "a dark Indian looking poor pennless Russian that have the name to be a writer" (28). As a writer herself, Harrison could see through the pretense.

"The American and English drink and gamble so much," she sweepingly charges, "the women after men the men though married are after other women those that are not during that are beating up the wife. or she are beating him up" (24). With some complacency, of the Roses she claims, "Of course they always talk to me as a maid and not as an American" (26). Her working-class identity eclipsed her nationality, rendering her a deracinated servant rather than a troublesome American. At least in France, to be American was to be licentious, profligate, and violent.

Showing that she took active steps to dissociate herself from her nationality, Harrison reveals, intriguingly enough, "I no longer own up to be American but are a Cuban" (24). She may have habitually represented herself as Cuban, given that three years earlier she had identified "Habana" as her home on a hotel register in Hawaii. In that instance, possibly she was motivated by a desire to forestall racist treatment, as her fellow guests were all white compatriots. In France, however, she alleges that she claimed to be Cuban in order to distance herself from misbehaving Yankees.

Setting up an implicit contrast to these dissipated folk, she follows a description of their excesses with an account of the "beautiful dream of the Most Holy Jesus" (24–25) that she had at the Roses: "I saw Him in my room under a horbor of ever greens and it was so plain I got up and called out Jesus" (25). However, closing some of the gap between them, she then changes course once again to enthusiastically describe the social activities of the circle with which she is affiliated if not a member: "The favorite gaiety among the American and English here are Surprise partys the Ladies bring some kind of cooked food and each man bring a bottle of liquire They arrive at 6.30 We had one here and wonderful good food. The Americans raved over my patato salad" (25). Her language leaves it open as to whether she attended the party as a servant or as a guest. Did she provide that "patato salad" due to the party rule stipulating that "the Ladies bring some kind of cooked food," or was it the fruit (or root) of paid labor? Even as she was serving these people, she seems to have shared in the festivities, as her pronouns shift from "they," to "we," to "my," to "you," and back to "they" again. "Everything are pyjmas," she comments, and she adopts the custom herself: "My Madame gave me a nice short coat to ware with pyjmas and it fit perfect" (25).

Harrison positions the local residents of "old Antibes" as the expatriates' wholesome foil. She states, "The people are nearly all working people and ever so family like. its 8 cafés where they meet dance sing laugh and talk" (26); "I have one homelike hang out its a real Family meeting place from the grandma to the children thats old enough to prance round on the floor" (27). Yet as a culturally fluid, highly mobile consumer, she had as much in common with the affluent outsiders she criticizes as with the French natives she admires. Moreover, in friction with its essentialist tendency to situate local populations outside of modernity, her narrative is peopled with numerous individuals with complex histories and subject positions. For example, despite her setting up other women as conservative cultural repositories, "flappers" make regular appearances in the text. Even at the homey café in Antibes, one of the managers was a "young Italian French lived in N.Y. very swell" (26) who liked to speak English with her. Her book confirms James Clifford's contention that by the early twentieth century, "local authenticities meet and merge in transient urban and suburban settings," forming a "truly global space of cultural connections and dissolutions."[44]

Finally, what to make of Harrison's taste for sunbathing? On her departure from France by train, she comments that a Dutch fellow passenger "were much amused to see me handle my two suite cases with so much ease that are all from those 100 of Sun baths at Juan-les-Pins. the Sea and the Pins are the thing" (268–69). The Murphys and their circle have a cultural footnote for triggering an American fad for such "baths." Quite suddenly, the dark skin caused by regular sun exposure came to signal health, wealth, athleticism, and status. Dubbing the practice "cosmopolitan tanning," Susan Keller explains that even though such complexions had historically been a condition of exclusion, in the 1920s "tanned skin became a key symbol of modernity . . . almost as important an emblem of the modern as the automobile."[45] Tanning expressed the racial confidence of a ruling class. Thus Fitzgerald used it to map the class geography of *Tender Is the Night*. On first venturing out to the beach, Rosemary, the novel's young protagonist, encounters "a group with flesh as white as her own" who "were obviously less indigenous to the place." Those sporting "very white teeth" and very dark skin were "mostly Americans, but something made them unlike the Americans she had known of late."[46] The girl's entrance into that rarefied society is marked by herself getting a tan.

Harrison embraced the trend in France, sunbathing if not deliberately tanning, another sign of her immersion in bohemian outdoor culture. Sunbathing, she contended, amplified her vigor. Her bold travels did the same, and they too darkened her skin. When she first returned to France from Egypt, she had commented on her new "Egypt and Indian brown" (177), adding—her satisfaction palpable—"I was a dark chocolate." Visible evidence of her intrepid travels, her Asian "brown" connoted vitality and worldliness, and she revived it on the Riviera.

There is a surprising instance in which Fitzgerald's and Harrison's narratives converge: In his Riviera essay Fitzgerald deploys the exact same adjectives as Harrison—"Egyptian" and "chocolate"—to characterize the skin tone he and Zelda acquired on the Cap d'Antibes:

> In the late afternoon of September 1, 1924, a distinguished-looking young man, accompanied by a young lady might have been seen lounging on a sandy beach in France. Both of them were burned to a deep chocolate brown, so that at first they seemed to be of Egyptian origin; but closer inspection showed that their faces had an Aryan cast and that their voices, when they spoke, had a faintly nasal, North American ring. Near them played a small black child with cotton-white hair who from time to time beat a tin spoon upon a pail and shouted, "Regardez-moi!" in no uncertain voice.[47]

Their daughter is still darker. Utterly adaptable due to her age, the "small black child with cotton-white hair" and the "no uncertain voice" with which she shouts commands in French has the ability to accommodate different subject positions, at once black, white, French, and American. The adults, however, can barely make a show of being anything other than what they are, with their American accents—and intractable American habits—giving them away.

In recording the speech of their server, Fitzgerald reveals that the man is actually a compatriot: "'The check, suh,' said a Senegalese waiter with an accent from well below the Mason-Dixon Line. 'That'll be ten francs fo' two glasses of beer.'"[48] The waiter's appearance in the text is brief yet telling. For one, it serves as further evidence of the expatriates' reliance on American amenities: crackers, newspapers, and a homegrown servant. That he should hail from the American South rather than France's West African colony is additional proof that the Fitzgeralds and their friends are not living a truly French life. But more important, in registering the presence of African American workers, the scene offers another example of how racial appearances and categories get transmuted on the Riviera. Harrison lived among other wayfarers who, just like her, could assume different ethnic and national guises.[49] Throughout the book she insists on her ability to simulate the local populations she moved through. Like the Americans whom Fitzgerald mocks, the distance she put between herself and her American identity was perhaps not as great as she liked to believe or at least profess. Nonetheless, her text proves that she went a considerable way.

Sharpley-Whiting comments that we have "expatriate yarns about everybody but African American women."[50] Fleshing out Harrison's snapshots of France, we can assemble a story about her time in that country if perhaps not an actual yarn. It is tempting to set up binaries between Harrison and her celebrated compatriots: Black/white, working/leisured, servant/artist, poor/rich. Her public and private writings, however, break down the distinction. Juxtaposing her with this community helps us better recognize Harrison's modernity, innovation, and investment in leisure, as well as her commitment to portraying it. Of the Mediterranean, Fitzgerald gushed, "It is a blue sea; or rather it is too blue for that hackneyed phrase which has described every muddy pool from pole to pole. It is the fairy blue of Maxfield Parrish's pictures; blue like blue books, blue oil, blue eyes."[51] He adds, "The sea and skies are running a race to see which can be the bluest" (FL2). Actually, though, that final sentence is not his, but Harrison's.

Her life in southern France was so felicitous that it may have quenched her thirst for travel, for an unusually long while at least. She likely would have stayed there much longer had it not been for the upheaval that convulsed Europe. As it was, she had to leave, propelled by economic forces rather than any immediate threat to her safety; the war to come was still not quite on the horizon. Although France's "relative self-sufficiency" had buffered it from the effects of the global Depression, in time it was overrun by galloping inflation.[52] As a consequence the tourism industry collapsed. From eight billion francs annually between 1926 and 1931, tourist expenditures shrank to just 750 million in 1935.[53] The precipitously overvalued franc scared away new visitors and propelled expatriates out of the country. Within Harrison's personal orbit, the Morrises left Paris to resettle in New York.

In line with her disinclination for political commentary, Harrison does not mention Germany's rising fascism or even voice a more generalized concern, whither Europe. She does note that she was unsure if the Roses would return to France from England, and she alludes to plummeting tourist numbers on the Cap d'Antibes: "There are a very few English and American Sunning this winter as in the pass. Well what was bad for them was good for me" (266). Typically, she figured the change as personal gain. However, by the summer she too was gone.

Later, in Waikiki, she often thought back to the sheer sensuous joy of her Riviera life. She happily concluded that her new home was "as much like Cap d'Antibes if it had been made to order" (312), and later chapters will discuss her experience there. But first, on to the circuit from northern Europe to southeastern Asia that she identified as the "last half of my world tour" (266).

CHAPTER 6

Imperial Geographies

My *Great, Wide, Beautiful World* was promoted as the outcome of Harrison's "ten-year plan to circle the world" and intention "to work her way around the world" (xii, x), an enterprise that commenced with an outgoing Atlantic voyage and concluded with a homeward-bound Pacific one. Marketing the book as a tale of circumnavigation capitalized on long-standing fascination with the feat, updating a flourishing genre of around-the-world narratives that originated in the New World discovery accounts of the 1500s.[1]

The emphasis on Harrison's intention "to circle the world" is in some respects deceptive, in that it frames her travels as breaking from her past activity and as proceeding deliberately onward in a single direction. Her world tour, as we know, was not a discrete trip but rather a continuation of an established lifestyle. Long before leaving North America, she had for years alternated work and travel. Moreover, once overseas she frequently doubled back to past destinations, and for half the book's time span she was stationary and settled in France.

Yet Harrison herself figured the undertaking as "my around the world trip" (165). From the onset, she had determined on circumnavigation, departing America on the Atlantic and returning on the Pacific. During her travels, moreover, the pursuit was central to her identity. She described her activity as "Globetrotting" and herself as "nothing but a glob trotter" (116, 294). Joyce Chapin identifies the early imperial assumptions that undergird such a perspective. Exhibiting "a proprietary confidence," to go around the world, she states, "was to demand membership in the club of nations that dominated the globe."[2] As we will see in the following pages, Harrison situated herself as an American woman, and sometimes as a European one, who had the right to engage in such a pursuit.

Were there any doubt, the title of her book insists on it with its commanding summons of "my world." At the same time, however, throughout the text her racial identity, servant's reality, trickster practice, and catholic social impulses continually disrupt any claims to supremacy. This chapter parses the privilege Harrison enjoyed as an American beneficiary of imperial legacies even as she had to leave her own country to access them. It pauses over selected experiences in Africa, India, and the Middle East, and it also tracks the last stage of her world journey, a rapid eight months of travel through northern Europe, Russia, and Asia.

In mulling over where she should eventually settle, Harrison once stated, "I am free to choose a place any part of the world."[3] Here she displays the cosmopolitan perspective that Valerie Popp characterizes as "the bedrock of 'global citizenship.'" "The global citizen constantly weighs multiple representative homes against each other," Popp explains, and Harrison's legal status as an American citizen facilitated such an outlook.[4] Innumerable destinations were opened up to her by their past or present rule under world powers that include Japan, France, Spain, Britain, and the United States. Japan aside, virtually all of the non-European destinations in her book were either former colonies or still under sovereign control. Absent such histories, locations including Hawaii, Somalia, Syria, Egypt, Sri Lanka, India, Hong Kong, Taiwan, and the Philippines, as well as the South American nations she toured in the 1940s, would have lacked the infrastructure that allowed ordinary Americans like her to readily visit them, including transportation networks, money exchanges, global-standard hotels, the widespread use of English, French, or Spanish, and the legion of unquantifiable perks granted Westerners. To put it another way, Harrison personally benefited from colonial regimes founded on the exploitation of native populations.

Yet due to both her race and class, her enterprise did not resemble that of the typical Western visitor. This African American tourist who funded her activities through service work carved out a very different kind of itinerary, by means of what Camilla Hawthorne and Jovan Scott Lewis eloquently characterize as the "situating force" of Blackness, "a place-making apparatus that in every geographic context makes its location more meaningful, more substantial, more human."[5] Reciprocally, the shifting political locations through which Harrison moved made her subjective locations all the more complex.

Such complexity is especially conspicuous in her encounters with African women, and from here I turn to zooming in on two of them. The first, previously glossed in the introduction, took place in 1929 at the Jardin d'Acclimatation in Paris, while the second occurred the following year during a port of call at Djibouti, then the capital of French Somalia. These interactions caused Harrison to devise her own "rhetoric of kinship" in positioning herself relative to the people she met, to use a term coined by Alasdair Pettinger in respect to African American responses to Africa.[6] Her experience corroborates Brent Hayes Edwards's contention, that in "transnational circuits, black modern expression" is formed "through the often uneasy encounters of peoples of African descent."[7]

Exhibitions of colonial peoples and their lifeways were a staple form of entertainment in France, self-congratulatory national statement that doubled as public amusement. The small-scale Jardin d'Acclimatation, located within the Bois de Boulogne in the western outskirts of Paris, launched the trend. The enterprise started out as a botanical garden and zoo that displayed plants and animals gathered from French colonies; during the 1870 Siege of Paris, its rare stock was served up to wealthy Parisians in dishes prepared by a celebrated chef. On reopening, it expanded to include living human exhibits, groups of African, Native American, and Inuit women and men occupying replicas of their native villages. The park became "the major French site for the exhibition of 'savages'"; for over half a century, its "collection of exotic animals and no less peculiar individuals" tapped into a voracious public appetite.[8] In all, thirty-four different groups of people were displayed there, under living conditions so poor as to prove fatal to some.[9]

1931 saw the Jardin's last such exhibit, featuring New Caledonia. This was the very year the practice culminated nationally in France's grand imperial display, the Paris Exposition, which ran for six months in the Bois de Vincennes. Fanning out across the French empire, government-backed collectors had brought back countless artifacts along with 1500 "indigènes" to "animate" the spectacle.[10] Of the experience offered by this "enormous panoply of exotic worlds," James Clifford narrates, "pavilions from all the colonies, costumes, statues, masks, curiosities of every sort, 'savage dances' regale the traveler in a land of well-ordered enchantment. Official marked paths lead the visitor without confusion from one outpost of progress to the next—Indochina, French West Africa, Madagascar, New Caledonia, Guinea, Martinique, Reunion."[11] Lynn Hudson puts it starkly: Such fairs "married the ideas of progress and white supremacy."[12]

The "village" that Harrison visited at the Jardin was inhabited by women of Central Africa, which was made up of French Gabon, French Congo, Ubangi-Shari, and French Chad. She passed an agreeable interlude with them as they conducted domestic tasks under the gaze of paying spectators, an encounter she memorializes in one potent paragraph:

> I climbed over the fence and got in the native village where the Plate mouthed women are. a slip is cut just wide enough in the lip to fit around the rim of a wooden plate they can hardly talk with it only the women have the plates there is about ten of them and they took a fancy to me. I think they saw I had some of their blood I couldnt fool them. the yongest wife was during the Cooking as I hung around the Camp fire she offered me some it was good and I would have accepted to save the price of my supper But the spit run out of her mouth on this plate and ofen droped into the pot. When I left I climbed over the fence again so it didn't cost me anything. (19)

The connections she makes are facilitated by not only her privilege as a Westerner but also her novelty as a Black traveler. That the women descry her African ancestry fans their warmth, she implies. Drawn to their fire, she transforms a scene staged for public entertainment into one affording private sociability, and her dodging the entrance fee further undermines the commercial premise. She is simply a hungry visitor hospitably received.

The episode, however, can be cast more darkly. The women must acquiesce to Harrison's voyeurism. Their only choice lies in the degree of welcome they extend. For Harrison's part, her appalled focus on their "plate mouths"—which she alleges hinders their speech—and rationale for rejecting their food spotlight the cultural gulf between them. The confinement of the ten women, near indistinguishable and collectively identified as "wives," sets off her license as a single American traveler and writer. The effect is magnified by her mode of exit, climbing over the fence that encloses them. Making sharper claims, Marina Magloire reads Harrison as positioning the Africans as enforced witnesses to her liberty, and she also mocks the indolence that checked her passing urge "to go way out there again" (20) to give them a coat she no longer needed.[13] Throughout the book, Harrison uses culturally traditional women—or at least those that strike her as such—to foreground her own free modernity, and this is an especially pronounced instance.

Harrison had an appetite for national celebrations and grand world's fairs, and she carefully planned her itineraries to satisfy it. "I made a good move when I chose Spain at this time of the Exposition" (174), she reports from Barcelona. Her narrative is studded with many such events, hosted by cities that also include Nice, Seville, Madrid, Brno, and Manila. Whetting that appetite, as a girl in Mississippi no doubt she had heard reports of the 1904 St. Louis World's Fair and longed to go; reminiscences about the earlier Chicago World's Fair surely also filtered back to her.

In some ways, this anthropological entertainment accorded with her guiding life principle: to assemble within her personal history a conglomeration of nations and cultures and to access all the forms of leisure, spectacle, and edification to be had from them. Harrison's internationalism was intertwined with an avid consumerism. Her world's fair visits appear as versions in miniature of her world journey—purchased with a ticket. Yet if the controlling impulse of these expositions was to situate all of humanity in its proper place, Harrison was invariably not in hers. Consequently, even as she haunted fairs, pageants, and carnivals, she disrupted their hierarchies and challenged their assumptions. As in her visit to the Jardin d'Acclimatation, she regularly closed at least some of the subjective distance between spectator and spectacle.

The same inclination to escape fixed structures is evident in her preference to work in expatriate households, to patronize ships with multinational passengers and crews, to linger wherever more than one language was spoken, to dwell in borderland regions. Of the Boulevard Marguerite-de-Rochechouart in Paris, she approvingly observes, "One step you feel in Russia next in Turkey" (18), and in Antibes, she remarks, "When I go out of the little street some say buno Sera Senerina I like being in a country and yet feel I am of another at the same time" (27). Her savor of contrast was boundless. As we saw in the last chapter, on the Riviera she hung out with both expatriate bohemians at their edgy parties and working-class French families at their homey cafes. Writing from Madrid, she enthuses, "I just love the way I am living on a job at the Hotel Savoy in the morden part of the City and through the day with American English French and Spanish. then at night I am over on my ancient old street with all Spanish" (217). She sought out cultural environments that, like herself, were "betwixt and between" (141), to borrow the assessment of a confounded admirer in Darjeeling, a "Burma lady" who recognized Harrison's cycling through class identities as an exercise of power and privilege. "One minute you have a fine time with the lowest cast," she charged, "next minute with the highest Hindoo. one minute you wear a blue suit next minute a dress of 2 cent a yard crape then a little velvet dress with diamond ear rings how can we tell. one minute you stay in a hotel at $5.00 a day then go to a resturant and have a 5 cent meal. youre betwixt and between" (141).

The following year, Harrison had another encounter with African women, this time in Djibouti. Similarly reverberating with colonial resonance, her excursion there reads like a protracted, less orchestrated version of her visit to the Jardin d'Acclimatation. Regulations at the French park had kept her experience predictable and her person safe—a sanitized dose of Africa—but her experience in an actual colony was less anodyne. Although she was protected by the authority of her French ship and the US consulate, this surety was at a distance, and she roamed the streets alone.

She made the Djibouti port of call in the course of a long voyage on the SS *Chantilly*, which took her in stages from Madras to Marseille. Her narration of a freak accident that occurred prior to departure sets the stage for the colonial dynamics of the Djibouti outing:

> A very sad accident happen just before we sail a young Indian man a passenger He was standing by my deck chair with his Parrot teasing a little French Girl. then he walked away and fell from the top Deck to the hull where they were loading bales of cotton he hit the third deck in the fall. If he live he will always be hepless they took him to the Hospital. When I looked down and saw the dead Parrot I was shocked for then I knew it was the young man that a half minute ago was so happy. It was beautiful to see what tender care the coolies with their red strip around their loins and a piece on their head handled the young man.... When they started off with the Boy on the stretcher one of the coolies put the dead Parrot by the Boy's side and the English Doctor threw it out But one of them Picked it up. (158–59)

The struggle between the Indian laborers and English doctor approaches slapstick, even as the "tender care" with which the former "handled the young man" casts in stark relief the latter's view of the parrot—and by extension, its owner—as literally disposable. The accident, moreover, is triggered by a male colonial subject's teasing of a small white girl. Were it fiction, readers might conclude that the scene demonstrates that any disruption of racial hierarchies will lead to deadly outcomes, offering a preview of the threat that Harrison's Djibouti trip posed.

Djibouti was the capital of a French colony whose history, 1884 to 1967, almost exactly corresponded with Harrison's lifespan. In 1896, following a series of treaties with the Issa and Afar people, the French designated it French Somaliland. A major port, Djibouti's population swelled with the completion of the Imperial Ethiopian Railway in 1917. Collectors for the Paris Exposition were likely active on the coast of Somalia just as Harrison was there.

In her account of her long day touring Djibouti, at every turn Harrison proves her sense of ease. Rather than any racial kinship, this comfort depends upon a ubiquitous state presence, which manifests itself in the institutions established by colonial authorities and managed by local workers. Harrison habitually identified more with the colonizer than with the colonized, and her perspective during her sole visit to Africa was no exception.

The outing commences with her taking a "native row boat" to shore rather than the motorboat the other passengers favor; she rather opaquely explains that she chose the rowboat "so to be out of style" (160). On arrival, she relays, "A native had just been arrested so that made it quite lively" (160). The unseen altercation, far from inspiring unease, only adds to the city's entertainment value. She likewise visits the police station in the role of spectator, "just to get a look at the Style of the men" (160). Regulating the local population to safeguard Westerners like her, the police as a bonus add to the show. She stresses that "Djibouti is French" (162), under the jurisdiction of her favorite nation, even as the presence of the US consulate makes for a sense of glad recognition. "I felt much at home on reaching the American Concel Building the Stars and Strips waving a welcome to me" (160), she remarks. Despite her sometime resistance to being identified as American, here she not only benefits from the perquisites of US citizenship but also feels an emotional response.

She continues on to a small park sporting a tennis court, a ground for white upper-class leisure that reads as the very mark of colonial occupation. Her status as a Western tourist prompts a "native boy" to offer her a "Kane settee" to rest on, an equally potent emblem of colonial ease. She perches there only briefly, finding the chair "to short" (160), with her physical discomfort perhaps reflecting her psychic discomfort about occupying an actual seat of privilege. Instead she "stretched out on the floor of the Band Stand," where she drifts off to sleep feeling absolutely secure.

After the "good nap" (160) she ventures away from these European spaces—the police station, consulate, and park—into Somali territory. She moves fluidly about the settlement, shifting between groups of women as she passes through the workers' district, visits the bazaar, pauses by the fountain, and seeks out glimpses of private homes. During this stage, however, conditions become more volatile. She threads her way through the "thousands of huts" that she dubs "the Coolies quarters," using the pejorative term for unskilled laborers employed by foreign corporate powers (161, 160). There she is approached by "several women and girls" who "came out and begin to dance" (160), soon joined by many more. In a rather ethnographic vein, Harrison observes, "Only a few ware ear rings nor anklets and few have their face cut with cast marks like you see in Egypt. Their orments are a long string of good beads and a bracelet on each arm above the elbores and a gold bead collor the women are all slender" (161–62). She also compares their customs with those of tribal women she had seen in Egypt to authoritatively pronounce, "Here are a mixture of Indian and Africans" (161).

"I suluted them in the Proper way and having a black Vail head dress very much like the married women ware they fell to me" (160–61), she alleges. Such a claim is characteristic of *My Great, Wide, Beautiful World*. Harrison often turns to other women to find cultural intimacy and confirm her status as a citizen of the world, efforts facilitated by her ambiguous ethnic presentation. Yet while she imagines that she approximates the Djibouti women's appearance and behavior, her text reveals that they view her as categorically exotic. She recounts, "In just a few seconds after suluting a few women I had swams around me first they thought I came from Greece after a while they decide I was Chinese as many as could get to me I had to shake hand it was very pleasant. then a woman came up to beg and another let her know she were not to beg me" (161). "Swarmed," even as Harrison registers the interaction as "very pleasant," one of them steps in to protect her.

"The bunch at the fountain" include a "little Flapper," whom the others expect her to recognize as a kindred spirit: "They said I took special notice of her because She had her hair bobbed and one felt my head to see if I had mine bobbed" (162). How did she know what was said? Were they speaking French together? Regardless, she does not take that "special notice." At least among strangers or at least in her book, other women best serve her as emblems of tradition, and she is uninterested in evaluating the younger woman's difference. "Flappers" nevertheless often crop up in her text. While Harrison usually presents other women as culturally static, a foil for her own innovation, the flapper's brief but regular showings indicate how frequently she encounters, although rarely dwells on, modern expressions of gender rebellion.

She depicts herself as engaged in a form of pursuit, explaining, "I would look down each street and if I see women I would go down that street" (161). At last she discovers the limit of their tolerance, and the tour culminates with her literally barred entry to their practice: "I meet a young woman in her finery and went along with her until we came to a house I could tell by the singing it was a new Bride's house when she went in I tried to slip in with her but She wouldnt let me and shut the gate to the high walled yard" (162).

On leaving the Somali women and their sometimes enjoyable, sometimes fraught, intercourse behind, Harrison's repairs to the western part of town, where she dallies past sunset in the "many side walk Cafes" (162) that catered to the settlement's French residents. The narrative then reverts to her experience in locally controlled space. As she was walking back to the ship, she recounts, "3 Boys cought up with me two of them begain to be fresh," with the euphemism "fresh" signaling sexual assault. She punches one of the assailants and leaves the other "running for life." Then "two others came up." One of them also "got fresh," and thus, "I landed one somewhere near his eyes such a groan and holding his face" (162).

Harrison looks to have been in real danger, surrounded by aggressive boys, or perhaps young men, alone and at night. However, she presents her assailants as ultimately the more fearful, due to not her physical might but rather the prospect of state repercussion for attacking a Westerner. Of the one among the five who did not seek to harm her, she states, he "tried to be nice and said I did right I could see that he felt uneasy walking by my side." Finally he removed his cap and bade her farewell in the colonizers' language, "Bon Soir Madame" (162). His deferential gesture and words—the sole direct quotation to appear in the Djibouti tale—confirm Harrison's affiliation with colonial occupiers. Confident in her status as the wronged party, she reported the attack to a man on the pier, perhaps a member of the ship's crew or a local authority. The man "spreaded the news" (162).

As a bookend to her tale, this dissemination of "news" of Native male trespass at the end of her outing joins the report of a Native man's arrest at its start. Folding the assault into a narrative of a pleasure trip, Harrison concludes by saying that she had had a "perfect day" (163). Her sense of personal security and traveler's gratification alike are left intact, even as her visit has unsettled at least one of the district's inhabitants. While the Djibouti excursion makes for an exceptionally rich scene, it is also typical of the book, in showing how comprehensively her gender, national, ethnic, and class locations dictated her experience.

From here I return to a more chronological account of her travels. Harrison left central Europe in September 1928 with a train out of Sofia. Entering the Middle East, she traveled in Turkey, Syria, Palestine, and Egypt. As she passed through French and British colonies past and present, imperial histories continued to serve her, most evidently in respect to the languages she could use. She does not, however, remark on the implication of the occasional request that she not speak French. Her inclination was to adopt the perspective of occupying authorities, as when she states in Palestine, "The laws are a bit better under the British but the Natives are getting not to fear them as much as in the beginning as they are not so strick I wonder why they Build so many fine Hospital instead of putting up factories so the many idle men can have work to do" (78). Deploying the term "the Natives," she positions herself on the other side.

Her dual guises as Western tourist and migrant worker converged while she relaxed in a park after touring the ancient capital of Heliopolis outside Cairo. She recalls, "An Arminin Nurse with a pretty little Girl call to me. She wanted to know if I would go to their house and do some washing and ironing I told her no but she thought I did not understand called to a young man and told him to ask me in French then she knew I understood Her at first" (78–79). These two nonwhite women, both in domestic service, communicated using the languages of imperial powers, English and French. "I understood Her at first," Harrison emphasizes. Yet it is unclear why the nurse viewed her as a candidate for this work. Was it because of the color of her skin? Even as Harrison assumed the subject position of an inheritor of colonial legacies, she was recognized as a member of a servant caste. In this instance, she resolved any potential identity schism by conceiving the prospective job as a missed tourist opportunity, remarking, "After she had gone I wished I had worked the half day just to see How much they paid and also for the experience" (79). Contrasting the prospect with more beguiling forms of recreation, she continues, "But you could not expect me after a good lunch with Icecream to top. to leave a nice shady Park. and go and do laundry work and it being my first trip to Heliopolis the City of the Sun."

Not incidentally, in Cairo she indulged her ice cream habit at one of the city's Groppi teahouses, another sign of her privilege. At this venerable Swiss-owned emporium—an icon of Egypt's "Belle Époque"—an elite clientele partook of gelato and chocolate while shaping the nation's future. The establishment is said to have "witnessed the politicians who were planning and starting revolutions, intellectuals who were discussing the future of the British occupation, and artists who were inspired" by its atmosphere.[14] Adel Toppozada, grandson of former Prime Minister Hussein Pasha Rushdy, reminisced, "We often went to Groppi's or Locke's, dressed up to the nines. These were real occasions and people always looked as though they were going to a party, the women in long evening dresses and fur stoles. Groppi's tea room was the place to people watch and be seen."¹⁵ Stepping outside any identity as servant, Harrison had both the right and the means to command service at this fashionable place, rubbing shoulders with Cairo's ruling class.

More definitive evidence of her social elevation in the Middle East, on leaving Europe she underwent a silent promotion, from nurse to governess. She discloses, "I refused several places as Governess in Egyptian Families because I do not like Little Girls of either Syria Palestine nor Egypte. I could not find a place with a little Boy" (82). The job title implies responsibility for educating one's charges, intriguing in light of Harrison's lack of formal education, which might not have been apparent to the Egyptians who petitioned her to teach their daughters. Her being an American and a native speaker of English sufficed to make her a desirable candidate. Harrison's longest position during her active travels was in Cairo, which she kept a full three and a half months, perhaps due to her relatively high status there.

Also of note is her admission that she didn't "like the Little Girls of either Syria Palestine nor Egypte." She does not give her reasons, but her antipathy may have been rooted in the conservative gender roles she encountered there. Her commentary on other women accelerated in the Middle East, and like many Western visitors, she fixated on the veil. While Harrison rather neutrally observes in Syria, "You see faces of beautiful women in the little norrow windows they stick their heads out and throw back their Vailes" (63), she offers stronger commentary in Istanbul: "You can see many women yet that ware the vaile I do not like to see them I feel like I are smouthering" (60). As usual, other women cast her independence in relief. Even as she represents the region's female denizens as especially constricted, she portrays herself as more robustly and even aggressively physical than ever—climbing a fence in Jerusalem to get a better view of a private home, hauling a girl onto a moving train in Syria, striking a man in Turkey who assaulted her. As a single traveler, she had to actively defend herself, and she positions herself as a rescuer of other people too.

We have seen that Harrison sometimes donned local clothing in an attempt to blend in on the streets. However, in more controlled settings she could do the same thing for the opposite reason, to attract the gaze of others—specifically, that of other Americans. She had a number of studio photographs taken of herself in Orientalist costumes. These she circulated among her friends: She jokingly proposed that she send a would-be employer "a picture with that Egyptian water jug on my head" (118), an image evidently known to her reader. The original dust jacket of *My Great, Wide, Beautiful World* features a collage of such photos, linking Harrison to a larger travel writing tradition in which authors model their cultural intimacy by posing in native dress. More metaphorically, her costuming resonates with W. E. B. Du Bois's famous contention in *The Souls of Black Folk* that African Americans, forced to see themselves "through the revelation of the other world," are "born with a veil."[16] Harrison put on actual veils to color how she, a Black female globetrotter, would be seen.

From Egypt she sailed to Sri Lanka. She spent a month on that lush rugged island, relishing its fiery food, beaches, mountains, and Catholic churches, along with the religiosity she encountered at every turn. Finally she reached India in April 1929, the terminus of the multicountry tour she had begun in Paris. After a short time in the port city of Madras, she proceeded to Mumbai, where exactly as planned she found tolerable jobs, including one as personal maid to a "quare" (120) woman wont to throw pillows at her and another as an attendant at a bathhouse. The latter catered to men who "just like nice brown mamas rubbing them up" (122), as she put it, her language leaving it open as to whether she included herself in that group.

Harrison's pleasure there and elsewhere depended on her ability to ignore volatile political conditions, both during her actual time in a place and in her written account of it. She "knew that the People all the way to Burmer are displeased and want home rule" (251), and she was intrigued enough by the independence movement to wish to spend more time in "Home rule" (107–8) communities. However, she does not ponder the motivations of resistant colonial subjects. "Rioton was very bad last week and I paid so little attention" (113), she reports from Mumbai. "They were expecting an attack then the troops were everywhere and I walking about sight seeing." Willfully ignoring politics, she narrows her purview to "sight seeing."

On leaving Mumbai, she commenced a vigorous five weeks of train travel. Stops included Kolkata, Agra, and Darjeeling. From the latter city, she remarked, "I wish that all the ones I love and all the ones I hate could see This Wonderful Place Darjeeling" (141). Thereupon she began an intrepid eastern journey on a route that seldom saw tourists, passing through present-day Bangladesh to Chittagong and then on to Yangon. Showing how carefully she devised—and communicated—her itinerary, a letter from the Dickinsons awaited her at *poste restante*. Thronging with migrants from across the Indian empire, in the late 1920s Yangon rivaled New York City as an immigrant port. Her comment, "what a joy to see the wide streets and the Women enjoying so much Freedom"

(149), captures its modernity. Originally she had planned to find work there. She changed her mind, however, turning down the interview requests prompted by the notice she had placed in the *Rangoon Times* to depart a week later for Madras—although not before sending some Burmese tea to Alice Foster in Los Angeles.

During her travels through the subcontinent, Harrison echoed many a Western traveler in exclaiming, "My how powerful rich one felt in India" (145). Now it was she who was in the position to dispense largesse to the help, as when she planned her grand exit from her hotel in Chittagong: "I will have a great time paying off the 5 or six men servents one cook for me another bring my meals another heat water, another bring up cold water for the bathroom another take out the night glass as they have no toilet another clean the room" (145). Yet, as usual, she destabilizes any clear-cut identifications. While she was far more able to purchase rooms, meals, tickets, and services in India than she was in other countries, it was also there that she commenced the most acute economizing on display in the book. Her practice hinged on its comprehensive railway system, which the British had assiduously developed to consolidate their rule. She bathed and slept in first-class waiting rooms and even spent the night in a carriage on a stalled train. Although such choices were far from the usual practice of a tourist, Harrison's privileged status as a Westerner gave her the assurance to make these bold moves.

Typical of her practice worldwide, when she had to pony up for a bed she relied on the YWCA. As a British colony, India had an especially well-developed network that catered to an Anglo-Indian clientele, and she patronized branches in Madras, Mumbai, and Kolkata. Staying at the Y gave her opportunities to meet fellow Christians in a nation where hers was the minority religion. It also introduced her to modern Indian women—living away from their families, working white-collar jobs, commuting by bicycle. YWCAs in India leaned upscale. Harrison was impressed by the lavish service at the Mumbai branch, "all men servants everything are served beautiful beautiful plates changed at each course. a tray in our room at 7.30 tea bread butter fruit then breakfast in the Dinning room at 9 then a lunch of 3 courses at 1 p.m. dinner at 8 and afternoon tea in our room at 4 p.m." (112). "You can see that a Christian home will soon empty your pockets" (112), she remarked, and she soon moved to a bare-bones apartment in a building that was once the annex of the bygone Watson's Hotel.

With this Harrison near literally entered Mumbai history. Opened in 1871, Watson's Hotel was a relic of the British Empire that in its heyday thronged with illustrious guests. As the city's first luxury hotel, it gave imposing form to the surge of wealth its cotton industry had unleashed. The black soil outside Mumbai was ideal for the crop, and trade burgeoned during the American Civil War after Confederate ports were blockaded. The influx of money went toward making "a grand showpiece" of its downtown core, with monumental public buildings of a style characterized as "Victorian Gothic."[17] Watson's designer, Rowland Mason Ordish, had done most of the drawings for London's Crystal Palace, which housed the Great Exhibition of 1851. If Harrison knew this, she was surely pleased, given her penchant for world's fairs.

Watson's Hotel was just around the corner from the Gateway of India, the imposing basalt and concrete arch monument erected to commemorate the 1924 visit of King George V and Queen Mary. Its floodlights made it Harrison's favorite spot to read at night. But unlike the Gateway, Watson's architecture spoke to not the past but the future. The multistory edifice was framed in thousands of cast-iron girders shipped in from England; anticipating the modern skyscraper, the iron frame remained exposed. A tourist attraction in its own right, it was described by one appalled observer as "like a huge birdcage exhaled from the earth."[18]

"The hotel of choice for colonialists and visiting dignitaries during the British Raj," Watson's sheltered both agents and chroniclers of imperialism.[19] Richard F. Burton, the British linguist and explorer, stayed there in 1876. So did David Kalākaua, the Kingdom of Hawaii's last king, to assess "the feasibility of populating his dominions" with Indian workers.[20] Mark Twain was also a guest, and his 1897 travelogue, *Following the Equator: A Journey Around the World*, describes the "shining and shifting spectacle" it presented. He imparts,

> India did not wait for morning, it began at the hotel—straight away. The lobbies and halls were full of turbaned, and fez'd and embroidered, cap'd, and barefooted, and cotton-clad dark natives, some of them rushing about, others at rest squatting, or sitting on the ground; some of them chattering with energy, others still and dreamy; in the dining-room every man's own private native servant standing behind his chair, and dressed for a part in the Arabian Nights.[21]

Twain's *Arabian Nights* pleasure notwithstanding, the hotel triggered a disturbing memory of how his father routinely "cuffed" the family's "harmless slave boy."[22] Seeing a German guest hit a worker, he writes, "carried me back to my boyhood, and flashed upon me the forgotten fact that this was the *usual* way of explaining one's desires to a slave."[23] The hotel makes a showing in Rudyard Kipling's short fiction, too: Contemplating whether to proposition a former suitor, the English protagonist of "Yoked with an Unbeliever" "sat for two months, alone in Watson's Hotel, elaborating this decision."[24] Not least it has a historical footnote as ground zero for India's massive film industry. On July 7, 1896, just one week after its first show in Paris, the hotel welcomed the Lumière Brothers' Cinematographe during its inaugural seven-city tour.

From her 1929 vantage point, Harrison described the edifice as "a handsome building that was the leading hotel 60 years ago and have several hundred one room apartments" (118). Watson's had long since been vanquished by its Indian competitor, the Taj Mahal Hotel, which offered a high-end alternative to the "Europeans only" institution so closely associated with the British empire. It too appears in the book, in Harrison's observation that "in just a block of here a great crowd stands every evening at the Taj Mahal Hotel under the ExKing of Afganistan Window until He comes and speak to them" (120). The "ExKing" was King Amanullah, in exile after abdicating.[25]

The building that once was Watson's still stands—albeit as a "decaying, crumbling mass of brick and iron."[26] Renamed Esplanade Mansion, the complex of small domiciles, businesses, and offices is especially popular with lawyers due to its proximity to the High Court of Bombay, "divided and subdivided into so many lawyer's cubicles to make them unrecognizable as the finest rooms in Asia that they were."[27] This "tiny cosmos ... where colonial heritage, film history and today's India meet"—to borrow copy from a recent documentary film—was made a UNESCO World Heritage Site in 2018.[28] At this point, it should come as no surprise that Harrison dwelled in such a fabled, fabulous place.

For Twain, Watson's was a site of Orientalist splendor and racial distress. For Harrison, the repurposed hotel was still another platform for her minimalist self-sufficiency. Commenting, "I think it just wonderful to live like this" (118), she minutely describes the simple yet optimal domestic appointments that allowed her to sleep, bathe, and cook for rupees a day. Yet the colonial past still resonated, and living "like this" connotes not only her spare lifestyle but also venerable social structures. Echoing Twain's emphasis on the abundance of "private native servants," she reports that the other residents in the building "felt so sorry for me the smartest looking servent boy came to me and said Mem Sahib which mean Lady I feel so sorry for you because you have no servent I will do what ever you ask me without any pay" (119–20).

Lest she be a figure of pity, a middle-class woman such as, putatively, Harrison, should have personal servants. She assumes the perspective of not labor but management in concluding, "This is one country where the servent problem never worry them at Bombay I did not have a Pillow to sleep on yet I had a boy servent" (145–46). The specificity of her language, here and elsewhere, also indicates her attention to, and apparent satisfaction in, the fact that many of the household servants—including the "nurse maids" (146)—were male. In India, she encountered service economies cast in novel forms.

Her revelation that she was addressed as "Mem Sahib which mean Lady" is telling. The Arabic term "Mem Sahib" originally meant the wife of a Sahib, or ruler; in India it came to mean the wife of a British official, in turn broadening to denote any white foreign woman of high social status. Once in India, Harrison became a "Lady," and she also began to refer to herself as a European. Whereas Du Bois maintained that "the black man is a person who must ride 'Jim Crow' in Georgia," Harrison suggests that the defining marker of the European woman was the expectation that she ride first-class in India, an expectation that included her. "European" historically connoted "white," but in India she too could be European. Class trumped ethnicity. Being in India actually made her darker. As noted in the previous chapter, she made a point of the "dark chocolate" skin tone she acquired there, her "Egypt and Indian brown" (177). Perhaps she savored the cachet of the holidaymaker's tan, or perhaps she was expressing a more profound satisfaction in her "cosmopolitan brown complexion," to quote Chimene Jackson.[29]

Her sense of elite membership, however, was tenuous. Suggesting its fragility, Harrison seems more prone to identify other people as white just as she begins to identify herself as European. A railway engineer, for example, is described as "a nice looking White Boy" (128). She also evinces a fascination with the "ladylike" (153) Anglo Indian women she scrutinizes, to the extent of commenting on their teeth and nails. At the Madras YWCA, setting up a racial contrast between one of the staff and "the Anglos" he waits on, she notes, "They have a butler with a clean white piece around his loins and a white head dress and his skin a clean shining chocolate color" (153). This observation reminds us of her own racial identity, which differs from that of both the man and the other guests.

Harrison exhibits, moreover, increasing resistance to the bounty of service at her command. Usually she loved immoderate attentions, such as those lavished on her by the Filipino cooks on a Chinese ship, "so delighted to serve me it is a free for all fight" (308). Yet in India she started to reject service, a response that suggests discomfort with occupying a position of privilege and dominance.

Her account of her stay at Agra includes an unsettling admission of inferiority: "I found a Good Hotel for $1.44 per day and no end of servients. to think an old waiter that served me had spent several years at London Nice and Italy in the service of Her most Highness Queen Victoria I felt so small to have him serving me. But He did it well" (132). The man reputedly once attended to the personal needs of the ultimate European, "Her most Highness Queen Victoria," ruler of the world's largest empire. Why did Harrison feel "so small"? And how to interpret, "But He did it well"? The language recollects her jolting statement about the rich young German woman who died in the Brno train accident: "Had I been killed it would have been absolutely nothing compared to that girl" (51). And finally: If a presumptive European identity makes one the natural recipient of service, then how did Harrison reconcile her own work as a servant in India, paid to meet other people's needs? Describing the nuts she sampled on an Indian train—kept hot in a pan "with something yellow on them and dry chilli" (124)—she makes a throwaway comment: "They can teach us how to serve nuts." Who is the "us" in this statement? Her text does not permit simple answers.

From India she went back to Europe for more travel and itinerant work, mainly in Spain (experiences discussed in chapter 9), followed by those four immersive years on the French Riviera. Her world tour recommenced in 1934. On receiving her visitor's permit to Russia, she left Paris to tour Scandinavia and visit the Soviet Union. In Moscow she boarded the Trans-Siberian Railway for China, finally realizing her desire to see the Middle Kingdom. Her tour of eastern and southeastern Asia included stops in Japan, Korea, Taiwan, Hong Kong, and the Philippines. By this stage she was no longer pausing to take jobs.

As we know, this extended journey took place after she and Morris had laid the groundwork for her book in Paris, arranging the travel letters retrieved from their original recipients. Thereafter Harrison had even greater motivation for continuing to write about her experiences, fleshing out the manuscript. One question is whether the book's set of final entries, too, originated in personal letters. I doubt it. Rather than routing them through intermediaries, it is more likely that she wrote travel accounts that from inception were slated for the book. Otherwise, she would have had to make further efforts to collect the latest missives. Victoria Lamont usefully describes textual production like hers as "a provisional mode of self-inscription, located somewhere on the continuum between private experience and its public circulation."[30] Once authorship hove into view, she must have written with all the more volubility and care. Indicating a heightened consciousness of herself as a writer with a quota to fill, Harrison commented of her experiences in Leningrad that she "could fill a hundred sheets" (277).

Her zest for novelty and edification continued unabated. She had long had Russia in her sights, if her studying Russian as early as 1930 is any indication. "I thought the world were small until I arrived here," she states from that country. "Leningrad is a world in its self I wouldnt give it for all the 25 or 30 countries and city that I have seen" (277). She also opines, "Its good to see it while its in the making. Here are no cast, and class no who's who. Here you are more free to do what you want to than any other country" (279). While the assessment, as Walsh points out, is painful to read given the looming Stalinist purge, her views align with those of internationalists such as W. E. B. Du Bois, Claude McKay, Eslanda and Paul Robeson, and Dorothy West. Tim Youngs explains, "Revolutionary Russia was a source of self-realization for many Black intellectuals and manual workers alike. In the USSR of the 1920s and 1930s . . . African Americans found social and economic opportunities that were denied them in the United States and claimed to experience an absence of racial prejudice."[31] Langston Hughes, for example, the restless poet who endorsed Harrison's book, found in the Soviet Union "political hope."[32] Similar to her compatriots, the enthusiasm Harrison voiced for the new order was also a critique of her home nation. At the same time, it is still another example of her contradictory attitudes toward social class. Her tutelage in Soviet egalitarian principles did nothing to stanch her relish of Western prerogatives in colonized nations or blunt her savor of proximity to wealth.

At the close of her eight-day train journey across Siberia, she alighted at Harbin, to then proceed south to Korea. "Fusan is French a Beautiful Port" (284), she rather inexplicably maintains. From there she embarked for Kyushu. Her merry account of the perilous voyage displays startling disregard—at least on a narrative level—for national trauma:

> We got the begining of the Typhoon that took so many lives next day at Osaka and the 8 hour crossing over to Shimonoseki were over or under the jolly Waves. they broke open the doubled doors and stair case of our floor. And we were saved by the quick work of puting wide planks and canvas sheets so when another big one came it was close. Never have I laughed so much. it were Japanese matting floors we sat or stretched out as we liked and when the waves tosted the Ship on the side about 300 Bright Kimonos women and children Babies men suite cases Tea Trays all went to one side nothing to hold on to. we was as helpless as though we had neither hands or feet it was just like you would sweep up a long row of bright apples in a trough then turn

it upside down again. the sliding was play to the children and They laughed to almost fits with me. (284–85)

Although she alludes to the "many lives" that were soon to be lost, the lexicon of play prevails: "Never have I laughed so much." Harrison barely acknowledges the passengers' grave danger, and one would never know from this passage that the "jolly Waves" that rocked the boat were the actual wake of a national tragedy, the Muroto typhoon. The typhoon was at the time second only to the 1923 Great Kantō earthquake as the biggest natural disaster in Japan's history. Over three thousand people were killed, and more than two hundred thousand were left homeless in Shikoku, Kobe, and Osaka; the latter's electric, water, and communication systems were virtually destroyed. By the time she wrote that entry from Kobe, Harrison had been plunged in the midst of the chaos. With its characteristic focus on sociability and pleasure, her narrative from the epicenter of disaster is so blithe as to feel near perverse.

She spent two months in Japan, and I will save her impressions of it for the next chapter, in the context of her relationship with her Japanese landlords in Honolulu. She then returned to China to visit Shanghai and Guangzhou, and from Hong Kong she took a Japanese steamer for a two-week holiday in Taipei, Taiwan—or Taihoku, Formosa, as it was known during the Japanese occupation, yet another colonized island that met her tourist needs. She spent a good portion of it prone on her "Palate" in her guesthouse, enjoying the meals brought to her by the obliging "Girl Maids" (307).

The Philippines were next. Harrison arrived in January of 1935 intent on "a rest of several months" (307) after traveling so actively since summer. That her passport was up for renewal may have contributed to her plan to linger. She could count on finding a full-service US embassy in that island nation, where three and a half centuries of Spanish colonialism were succeeded by two transformative decades of American occupation. The United States had defeated Spain in 1898 in the Spanish-American War, and the Philippines remained under de facto American control until the mid-twentieth century.

She made sure to arrive in time for the Manila Carnival. Organized by the American administration, the purported goal of the "greatest annual event in the Orient," as it promoted itself, was to celebrate Philippine-American relations and economic development.[33] Being there surely prompted memories of her time in Hawaii, with its force of Filipino laborers. Although Harrison often depicts Filipino migrant workers as mistreated, her remark about Filipina women, "the nicest that you can find in any Part of the World," reflects her usual approval of colonial processes: "Their education are Foriegn with other nations govering" (308). Allowing that she spent a "delightful month" in Manila, she continues, "but I wouldnt care to live there" (308), without further elaboration. She does, however, describe a two-week "Southern Cross Cruise" (308) through the archipelago, a tourist staple in the 1930s.

While still in Shanghai, Harrison had remarked, "I want to stretch my cash to take me to Singapore Java and Baile then back to Honolulu" (293). Her itinerary also included "Siam." The plan was to tour Southeast Asia until she had nothing left for boat fare. Yet as it turned out, Manila was her last stop of any length before returning to US territory, due to her failure to procure affordable boat tickets to Indonesia and Thailand. She explains that in Asia, "The English and all the other European companys make a rule to let no European woman Travel 3rd and often there are only 1st class" (294). While she circumvented the policy in Shanghai by securing a berth on a Chinese steamer, in Manila she discovered that the occidental shipping companies "sell only 1st Class to European women and there were no oriental line" (309). Being American made her a "European woman," and so the policy applied to her, too, a strange reversal for a woman who was once restricted to the inferior carriages of Jim Crow. Were she a man, Harrison indicates, she might have had other options, but as a Western woman she was shut out of budget travel. The policy diminished her travel satisfaction as much as her pocketbook. She clarifies, "I wouldnt waist my money traveling with other Europeans" (294).

When the time finally came for her multistage return voyage to Honolulu and the close, by one measure, of her world journey, Harrison was well served by the transpacific trade network that generations of imperial enterprise had laid. She left the Philippines for Hong Kong in early April, passed through Shanghai once again, and from there continued on to Kobe. She had bought a through ticket so as not to exhaust her funds, her decision to forego a longer stay in Japan further motivated by "it being a little late for the Cherry Blossom" (308).

After the ship docked in Kobe, she made a hurried trip to Tokyo to see the cherry trees in Ueno Park. Her final dispatch from Japan: "The Blossoms was still beautiful and I was in hopes I might miss the Boat" (309). While one should take with a grain of salt her claim that she would not have minded missing the boat—losing both her luggage and her fare—the statement does suggest ambivalence about leaving Asia and perhaps also about going "home." In Kobe she boarded the SS *President Lincoln* for the last leg of her journey.

PART III

After *My Great, Wide, Beautiful World*

Celebrity and Beyond

CHAPTER 7

Hawaii Interludes

Ten days after leaving Kobe, the SS *President Lincoln* sailed into Honolulu Harbor. Harrison would have seen a skyline far more pronounced than that evident on her voyage a decade earlier, with the 184-foot Aloha Tower rising above the modern buildings to which the once humble jumble of wooden homes and shops had given way. With supreme satisfaction, she concluded, "Well Wed April 24 will end my Plannes of 10 years ago and They have been far more beautiful and plesanter than I had any thought of" (310). She was on the verge of authorship, and the prospect of publishing her travelogue in American presses may have helped her imagine herself back into American territory.

Harrison's publishers announced the centrality of Hawaii to her life and book by promoting *My Great, Wide, Beautiful World* with a photograph of her walking on a Waikiki beach (the same photo featured on this book's cover). Hawaii was her cornerstone. Between 1925 and 1967 she lived in Honolulu for three separate intervals, a span of time during which the islands were transformed: from an American territory to an American state, from a plantation economy to the brink of a tourist economy, from a white supremacist oligarchy to an exemplar of interethnic alliance. Harrison's life, too, changed dramatically. Her first visit was also her first venture outside of North America. By her last return she had lived on or traveled through four continents.

In the winter of 1925, Harrison embarked in Los Angeles for six months in Honolulu. In the mid-1930s, she suspended her world journey there, making it her close to five-year resting place before she moved on to South America. Sometime after Hawaii became a state in 1959, she went back, and she is buried on Oahu's windward north shore.

Hawaii played a special role in her life. It was the hinge of her world travels—bridging Asia and South America—and ultimately her final home. "I feel very happy to have choosed this lovely part of the World" (314), she affirmed. It has an equally key role in *My Great, Wide, Beautiful World*, in that her Pacific return to American territory yielded both a circumnavigation plot and a potent epilogue. It seems truly poetic that her book should end with her in Waikiki.

The 1930s Honolulu residency is by far the most represented period in Harrison's history, since it coincided with her authorship and celebrity. The final pages of *My Great, Wide, Beautiful World* describe its onset, a narrative that a number of her personal letters extend. Local public records document key facts. Honolulu newspaper profiles on Harrison and articles about visits from notables such as Ellery Sedgwick and William Pickens provide further insight, while the photographs that accompanied many of them contribute an important visual record. Such a wealth of material allows for an especially complete investigation of this time. Thus it is treated in not only this chapter but also the next one, which focuses on the reception of *My Great, Wide, Beautiful World*.

Harrison's 1925 Honolulu sojourn closely followed on the bank failure in Denver that torched her savings. In the book's preface, Morris maintains that the fortunate outcome of this crisis was that it led her to Los Angeles and the Dickinsons. She overlooks the fact that the loss brought her to Hawaii as well. The Port of Los Angeles had recently become a point of embarkation for regular sailings to Honolulu, transporting not only laborers but also an increasing number of tourists. In the vanguard of holidaymakers enjoying Hawaii's increasingly upscale tourist facilities, the Dickinsons had made a very early trip to Hawaii in 1910, sailing on the Pacific mail steamship *Siberia*; in the early 1920s, they went twice more.[1] Their Pacific tales might have fanned Harrison's desire to go. More materially, they demonstrated how easy it was to reach Honolulu from Los Angeles—just hop on a boat.

The territory's tourism industry had entered a growth spurt. Like many liners, Harrison's first vessel, the SS *Calawaii*, was repurposed for leisure travel after the number of immigrants entering the United States plunged, due first to World War I and then to stringent national quotas. Having once run transatlantic routes as *The Sherman*, bringing European immigrants to the United States, the ship had been refurbished by the Los Angeles Shipbuilding and Dry Dock Company to ply the Pacific under a fitting new name.[2] Although its maiden voyage was just the previous year, *Calawaii* was already a popular liner, with the capacity to transport 178 first-class and fifty-two third-class passengers.

Harrison sailed third-class, opting for the cheapest fare that was sometimes denied her, as a Western woman, on European shipping lines in Asia. That her ship was called *Calawaii* seems apt, as both California and Hawaii became her main American bases, linked by a Pacific voyage.[3]

By then, the islands had undergone more than a century of sweeping change. Following the voyages of Captain Cook and his crew, the first Europeans to reach them, they became a magnet for whalers, traders, and missionaries. The Kingdom of Hawaii was founded in 1795 when most of the islands were unified under one government, which soon expanded to encompass the entire archipelago. Its citizens, however, were ravaged by the infectious diseases that the new trade networks carried. Estimates vary, but while native Hawaiians numbered in the hundreds of thousands on Cook's first arrival, by the mid-nineteenth century the figure was scarcely seventy thousand.[4] In 1893, American colonists deposed the royal family to found the Republic of Hawaii, an effective coup that paved the way for annexation by the United States five years later. Greed, imperialist ambition, and white supremacist convictions drove the occupation, with the Spanish-American war providing the pretext for establishing an American stronghold in the Pacific.

Despite the native Hawaiians' decline, by 1916 the islands' population had more than doubled, their ethnic makeup permanently transformed. In her book, Harrison portrays Hawaii as a smorgasbord of delectable cultural events, including "Chinese gatherns, Japanese, then Hawaiians and when I feel Spanish There are the Potegeise" (314). But this assembly of cultures had not only a dark past but also a vexed present. Seized from a depleted Indigenous population, the territory developed an economy and entire social structure predicated on exploiting immigrant workers. A "hierarchical, paternalistic plantation society" was "firmly ruled" by a small white minority, largely the descendants of Protestant missionaries.[5] A group of family-owned corporations, colloquially called the "Big Five," controlled 90 percent of the economy. The primary driver was sugar, which had entered large-scale production following the reciprocity treaty of 1875 that eliminated US import taxes and allowed planters to hire foreign laborers on multiyear contracts. Annexation freed these workers from their bonds, further stimulating expansion. Planters aggressively recruited in not only China, Japan, and Portugal, which supplied the largest numbers, but also in Korea, Puerto Rico, Spain, England, Germany, and Russia. After the Exclusion Act of 1882 shut down Chinese immigration, the sugar industry relied on the Japanese, who became the islands' largest demographic. In turn, when Japanese workers demanded better terms, Filipinos were recruited to undermine them.

Contemporaries decried Hawaii's "system of peonage."[6] Different nationalities were pitted against each other to keep wages cruelly low. Nonetheless, many Asian workers had the skills and resources to cross over from toiling on plantations to running farms or small businesses of their own. Their success led white residents to fear them as an economic threat. Hawaiian men, on the other hand, while not viewed as competition were sometimes conceived as Black. That African Americans were so few in Hawaii made Blackness a more portable ethnic category.[7]

Harrison's first visit came on the heels of the Filipino Piecemeal Sugar Strike, a protest that ended in a deadly armed conflict. Eventually Japanese and Filipino laborers formed a pathbreaking interethnic alliance to strike against the Hawaii Sugar Planters' Association. In 1932 the "Ala Moana trial" made for another kind of pathway toward social reform, even as it exhibited the islands' systemic racism. Thalia Massie, a young naval wife, claimed to have been sexually assaulted by a group of men that included two Native Hawaiians, two Japanese, and a Chinese Hawaiian. All were acquitted, as there was a complete lack of evidence linking them to the crime. Yet in a shocking turn, Massie's husband and mother turned vigilante, kidnapping two of the accused and murdering one of them, Joe Kahahawai. They were found guilty, but the judge commuted their ten-year sentence to a single hour. Nevertheless, David E. Stannard argues, the killers' guilty verdict marks a turning point in Hawaii's history. Over the next decades, the islands effected an "astonishing reversal," in his words, from the white supremacist oligarchy of the 1920s and 1930s to a liberal, progressive society committed to equality in principle if not always in practice.[8] In 1963, President John F. Kennedy could describe Hawaii as "what the rest of the world is striving to become," a transformation that made the islands an even more agreeable home for Harrison.[9]

The territory's racial tensions did nothing to disrupt the accelerating American vogue for all things Hawaiian. A cluster of corporate entities fueled the illusion of a Polynesian idyll, chief among them "tourist boards, steamship lines, sugar and pineapple advertising agencies, and the mainland motion-picture industry."[10] To take one odd initiative, the Hawaiian Pineapple Company, now Dole, sponsored a nine-week visit by Georgia O'Keeffe to produce artwork for its advertisements; to their consternation, she neglected to paint a pineapple.[11] In the 1920s, Hawaii vacations were still the province of upper-class people like the Dickinsons and adventurers like Harrison. Yet even so, visitor numbers soared. The Matson Navigation Company's launching of the SS *Malolo* initiated regular steamship service to the islands. Matson also erected the Royal Hawaiian Hotel and bought out the Moana Hotel, once Hawaii's sole luxury accommodation. Its hotels put up the visitors that its ships brought in, a remunerative equation. Blazingly pink and of a fanciful Spanish-Moorish design, the Royal Hawaiian Hotel made Honolulu a resort destination that could entice a well-heeled clientele. As we shall see, it featured in Harrison's time there.

The city's upmarket facilities were concentrated along the turquoise waters and white beaches of Waikiki. Once separated from the city of Honolulu by an expanse of wetlands, Waikiki had been a Royal Retreat for nearly two centuries, rejuvenating rulers right up to Queen Lili'uokalani, Hawaii's last sovereign monarch. During Harrison's first visit, the controversial Ala Wai Canal project was under construction. Draining the marsh, the government evicted the mostly Asian small landowners who had managed its profusion of fishponds, duck ponds, rice paddies, and taro patches. Waikiki was transformed from an agricultural backwater into a playground for an international elite.[12]

Tourists enjoyed traditional dance, music, art, and textiles. "Beach boys" were a special draw, the expert watermen working out of the Moana and the Royal Hawaiian who purveyed Hawaii's low-key hospitality ethos while introducing visitors to water sports like surfing and canoeing. Don Blanding, celebrated author of poems such as the prurient "The Virgin of Waikiki" and the racist (but long beloved) "Vagabond's House," described a densely populated resort space that he dubbed "Flappers' Acre," where jazz bands "fling dance tunes" and "automobiles swish in and out of the narrow streets, filled with pretty girls."[13]

W. Somerset Maugham's 1921 short story "Honolulu" profiles the rapidly modernizing city that Harrison encountered. Although his racial caricatures are discomfiting, Maugham does convey its energy, variety, and paradox:

> Nothing had prepared me for Honolulu. It is so far away from Europe, it is reached after so long a journey from San Francisco, so strange and so charming associations are attached to the name, that at first I could hardly believe my eyes.... It is a typical western city. Shacks are cheek by jowl with stone mansions; dilapidated frame houses stand next door to smart stores with plate glass windows; electric cars rumble noisily along the streets; and motors, Fords, Buicks, Packards, line the pavement. The shops are filled with all the necessities of American civilization. Every third house is a bank and every fifth the agency of a steamship company.

> Along the streets crowd an unimaginable assortment of people. The Americans, ignoring the climate, wear black coats and high, starched collars, straw hats, soft hats, and bowlers. The Kanakas, pale brown, with crisp hair, have nothing on but a shirt and a pair of trousers; but the half-breeds are very smart with flaring ties and patent-leather boots. The Japanese, with their obsequious smile, are neat and trim in white duck, while their women walk a step or two behind them, in native dress, with a baby on their backs. The Japanese children, in bright coloured frocks, their little heads shaven, look like quaint dolls. Then there are the Chinese.... Lastly there are the Filipinos, the men in huge straw hats, the women in bright yellow muslin with great puffed sleeves.[14]

From the perspective of this visiting Englishman, Hawaii was exotic but recognizably "Western" nevertheless. Such portraits contributed to its reputation as an enchanting tropical outpost that could still offer the familiar comforts of home.

If she kept to her usual habits, upon first landing in 1925 Harrison found short-term lodging at YWCA O'ahu, which at the time was located close to downtown in the gracious Fernhurst Building, known for its gardens. On her return in 1935, we know that she did indeed stay at the Y's next location. Like its sister branches worldwide, YWCA O'ahu was dedicated to providing working women with affordable housing along with educational programs and job placement services.

For her job search, options included the YWCA employment agency and newspaper Help Wanted ads. She surely found work quickly, given the recent implementation of the Immigration Act of 1924, which virtually halted Asian migration to the United States and its territories. The abrupt exclusion of the Chinese and Japanese workers who had dominated Hawaii's labor market created opportunities for other people. Although Harrison claimed she could get a job whenever and wherever she wanted, arriving during an employment crisis could not have hurt.

Toward the end of that six-month interlude, she spent a single night, May 27, 1925, at the fabled Volcano House on the island of Hawai'i in Hawai'i Volcanoes National Park. The hotel offered views of Kīlauea, the thrillingly active volcano that had been a major tourist attraction since the mid-nineteenth century. It was advertised thus in a 1912 issue of the *San Francisco Call*: "Situated on the brink of Kiluea Crater, overlooking the molten lake of fire. Four thousand feet above sea level, affording a sublime view of the snow-capped mountains Mauna Kea and Mauna Loa. Every comfort that one would expect to find a first-class resort [sic]. The hotel is well managed, has an excellent chef and is thoroughly comfortable in every sense of the word. THROUGHOUT THE YEAR."[15] During her short visit, Harrison walked up to the popular lookout point and perhaps also made a longer hike down to the crater.

Jack London had stayed at Volcano House some twenty years prior. Mark Twain had done so even earlier, when it was still "a primitive hostel," to use his phrase. (As we know, Harrison would continue to follow in Twain's footsteps on renting a room in the complex that was once Mumbai's legendary Watson's Hotel.) He describes the spectacle in "Letters from Hawaii," the travel dispatches commissioned by a Sacramento newspaper that were later incorporated into *Roughing It*: "The greater part of the vast floor of the desert under us was as black as ink... but over a mile square of it was ringed and streaked and striped with a thousand branching streams of liquid and gorgeously brilliant fire! It looked like a colossal railroad map of the State of Massachusetts done in chain lightning on a midnight sky. Imagine a coal-black sky shivered into a tangled network of angry fire!"[16] Harrison surely wrote about the awesome sight with equal enthusiasm, and perhaps with equally arresting metaphor.

To reach Volcano House, visitors had first to board a ferry to the island of Hawai'i and then travel on its railway, whose construction in 1901 galvanized park visitation. The hotel's location in the island's rugged interior made for an especially insular tourist experience, with guests in close contact and doing the same activities.

We know exactly who was at Volcano House with Harrison, as each week its guests were listed in the *Honolulu Star-Bulletin*.[17] Judging by their names—the likes of Miss Frances Green, Mrs. Arthur Hepburn, and Mrs. Wilson Brown—they were a highly Anglo assembly. Many were officers' wives, visiting Hawaii while the men conducted training exercises with the United States Navy's Pacific contingent. They had traveled from their home states to see off their husbands and enjoy a tropical vacation footed by the government. On their arrival in Honolulu, the American Legion Auxiliary feted them at an afternoon reception—tea was poured and iced coffee served (the latter by none other than Mrs. Walter F. Dillingham, wife of the Honolulu oligarch).[18] Now they were off on a tour. The group included Mrs. Robert E. Coontz and Bertha Coontz, wife and daughter of the celebrated Admiral Robert Coontz. Commander-in-chief of the United States Fleet, Coontz was soon to lead the Pacific contingent on a goodwill tour to New Zealand and Australia; at fifty-six battleships, it was the largest armada ever to dock in the latter country.[19] Remarkably, Harrison, just off her most recent job as a maid, vacationed alongside the family of this enormously powerful man.

The National Park Service has preserved the Volcano House's visitor logs. The May 1925 guestbook attests, in Harrison's distinctive round hand, to the presence of one "Mrs. Juanita Harrison, Habana, Cuba." These five words make for a twofold subterfuge. What's with the Mrs.? And more important, why Cuba?

Although it makes no showing in her personal correspondence, Harrison sometimes used the title of "Mrs." to up her respectability, both when putting her name to public documents and when writing to strangers. That many of the hotel guests were traveling in the official capacity of *wife* may have been further motivation to present herself as hitched.

Her claim to Cuba is less easy to account for. She had left the Caribbean island at least three years earlier, in the interim living in the American South, Midwest, and West. The passenger manifest of her outbound voyage from Los Angeles to Honolulu gives Denver as her place of residence.

The reputed "Paris of the West Indies," in the 1920s the Cuban capital had a glamorous reputation. By naming "Habana" as her home—with the Spanish spelling confirming her insider status—Harrison could borrow some glory. Even more intriguing is the prospect that she might have claimed to actually be Cuban. Thanks to her long residence there she spoke good Spanish, and the odds were low that the other guests knew the island. Appearing to be Cuban rather than African American decreased the risk of being refused lodging and fostered more cordial interactions with prejudiced white Americans. Some of her contemporaries did indeed represent themselves as Latin American for these reasons. James Weldon Johnson reveals in his autobiography that on occasion, as Joseph T. Skerrett Jr. relays, "his ability with idiomatic Spanish allowed him to 'pass' for some other kind of brown-skinned man than an American Negro."[20] As we know, Harrison admitted to pretending to be Cuban while in France some four years later. Reporting on the deception in her book, she gives a national, not racial, rationale: "I no longer own up to be American but are a Cuban. as the American and English drink and gamble so much" (24). However, in Hawaii, surrounded by people who were themselves all American, her motive must have been different.

She sailed back to San Francisco in July. If she followed her usual pattern, she had used most of her savings for the holiday she took after quitting whatever job she had had. Living in the territory had tutored her in finding cheaper passage: Whereas the passengers on her outbound ship included many Anglo tourists, on her return the majority were Filipino laborers, and she was one of the few US citizens aboard born outside the Philippines.[21] She continued on to Los Angeles, where she joined, or rejoined, the Dickinson household. Yet those six months had made such a strong impression that ten years later she would elect Hawaii as the site of a permanent home—or at least what counted as permanent for her. As her ship approached Honolulu Harbor in April 1935, she had mused in respect to her savings, "Perhaps with what I have left I can begin paying on a little future nest in Hawaii and it will be a pleasure to look forward just as it have been a pleasure traveling. I hope it will come as easy as my traveling have but if it gives me any drudgery or displeasure then I do not want a nest" (311). Intriguingly, even as she was on the brink of reentering Hawaii, she envisioned the small home she might purchase there only as a "future" residence.

In true jet-setter fashion, during her world trip Harrison had chased the sun, culminating in a resolution to settle "where there are always Spring" (279). Los Angeles was never a contender, despite her self-identification as a Californian. While she did consider Nice, in her view its climate, like that of Los Angeles, was subpar, and during her final residence there she determined on Honolulu. Writing from the Mediterranean city, she explained, "If I had not a more delightful country and place [Hawaii] already selected to settle down in when I am tired of roveing about I would choose Nice but like Los Angeles it's not up to my ideal because of the unpleasant winter days. But of the two I prefer Nice. As I am free to choose a place any part of the world I was careful in selecting a place delightful the year around and not too hot in summer."[22] The attention to weather might cause one to overlook the redoubtable cosmopolitanism she expresses: "I am free to choose a place any part of the world."

Waikiki had all that she liked: sun, ocean, gorgeous scenery, flexible work, outdoor leisure, cultural diversity, and domestic options. Unusual for a resort location, it also had an "urban soul," as Alika Bourgette formulates it.[23] The pastoral in the city was highly attractive to Harrison, who throughout her life sought out scenic locations—especially warm ones by the sea—even as she also gravitated to world capitals. Crucially, that it was a US territory and, later, a state, meant she could live there as long as she liked, with the perquisites of American citizenship. Moreover, despite its Depression-era economy and distinct labor challenges, she could find very good jobs. As a bonus, by all evidence she had entrée into numerous social circles, from the working class to the elite. 1930s

Hawaii had an expanding upper class, large military population, international labor force, and fleet of European artisans. With her inveterate crossing of class, ethnic, national, and cultural lines, Harrison consorted with them all.

Even as political strife precipitated her departure from Europe, 1935 was far from an ideal time to return to American soil. The economy's recovery from the Great Depression was in its early stages, and it would not fully rally until the Second World War. Hawaii's sugar and pineapple industries remained well below the peak levels of cultivated acreage and production they had reached in 1932.[24] In December 1936, close to two years after Harrison's arrival, roughly one-quarter of Hawaii's labor force was still unemployed.

The stalled economy made for static demographics. In contrast to Los Angeles's rapidly growing Black population, Hawaii's had for decades barely budged. The territory had 348 African American residents in 1920 out of a total population of 255,912, and 563 out of 368,336 in 1930.[25] A decade later—just after Harrison's second departure from Honolulu—the population was at its lowest point, a scant 255 women and men. African Americans thus lacked the cultural influence and lobbying power of larger groups. However, their low numbers also meant they had no clear position within the islands' ethnic hierarchy, and as a result they suffered less systemic racism. Barack Obama's statement about the Hawaii of his youth in the 1960s is even more apt for that of the 1930s: "There were too many races, with power among them too diffuse, to impose the mainland's rigid caste system."[26] The ethnic tensions and strictures were of a different kind, centered on Asian and native Hawaiian populations.

The islands' ethnic diversity, lack of Black-white binaries, and, especially, controlling presence of Hawaiian culture created an environment in which Harrison could experience different conceptions of race. "The Pacific offers expansive ways of being and belonging that embrace, cross, and exceed 'race' and 'indigeneity,'" Nitasha Tamar Sharma argues in *Hawai'i Is My Haven: Race and Indigeneity in the Black Pacific*.[27] "Coming to a place that outlawed slavery, offered citizenship, and viewed darkness as powerful, members of the African diaspora have long found the Islands a place of increased opportunities. . . . Hawai'i tempers White supremacist ideologies because of the survivance of Hawaiian ideologies." Although Hawaii was far from the social paradise touted by the era's tourist industry and sociologists alike,[28] making it her home gave Harrison some distance from the discrimination and prejudice that African Americans endured in the United States. It also, to quote Sharma again, gave "sanctuary from strict and seemingly fixed and authenticating ideas of Blackness that are sedimented in U.S. history."[29] As so often in her life, in Hawaii she was a minority of one, with fluid, multiple affiliations.

Unlike the rest of *My Great, Wide, Beautiful World*, the epilogue about her Hawaii experience does not have date and place headings. Instead, the single long entry that comprises it is headed "Now." Harrison thereby positions the Waikiki sojourn in a timeless utopic present. The epilogue's claim to "now" is anticipated in the two sentences that immediately precede it, which were deemed eloquent enough to close the second *Atlantic Monthly* excerpt: "I can swim and have the sand and large brown rocks and the long stretch of cocoanuts trees. as I write this my mind travelle to Zeurich Switzerlan Zufolo Iseral Philistine Sweden and Madras, Ceylon and Smiling faces stands out before me. I say good morning and eat a nice Pineapple the sweetest that grown The Birds are awaken I'll soon have to dress" (314). This instance is one of the few in the book in which Harrison stages her actual writing, leaving her perpetually in Waikiki, perpetually pen in hand at the start of another blissful day. The tropical setting, coupled with the early hour and the act of writing itself, triggers a lyrical fugue state, an evocative reverie in which a myriad of places and people dreamily fuse.

Note her emphasis on possession: "I can . . . *have*" the sand and rocks and those cocoanuts trees—referring back to the "my" of the book's title, with its imperial assertion. Indisputably, Harrison contributes to a problematic tradition of representing Hawaii as paradise, ripe for the outsider's delight. As a whole, however, *My Great, Wide, Beautiful World* posits that years of domestic service twinned with deep cultural exploration have earned her the right to at least temporarily enjoy this happy interlude. Her provisional settling in Waikiki both closes her travel narrative and promises a sequel. She has paused but not stopped.

The book also offers a good deal of information about the material conditions of her new life, and how she established them. Harrison kicked off her renewed Oahu residence with a five-day pleasure fest. A highlight was a gathering whose attendees, she recounts, "had Hulu dancing by the best singers and dancers then a Hawaiian feed Barbecure pigs steamed sweet potatoes cup cakes and punch. afternoon Baseball then a great time with singing contest we laughed so head" (311). Her insistence that the event was a "real Hawaiian Picnice" only foregrounds its identity as the opposite, a once customary occasion monetized for paying tourists. Harrison was always ready to sample other cultures in packaged forms.

The day before reaching Honolulu, she had predicted that since she had "not much left" of her savings, she would "have to go out to get a job though against my will" (311). Two weeks later, echoing many a new arrival to Hawaii, she reported that she "found the room rent and food very high," and so, sure enough, "went out to get a job although against my will." A reproving piece in the almanac *Hawaiian Annual*, "Popular Misconceptions About Hawaii," describes the starry-eyed newcomers she found herself among. Well conveying both Hawaii's romantic reputation and the corporate campaigns that disseminated it, its author, Will Sabin, maintains that the flocks of would-be islanders had been dangerously misled. The dreamy Pacific ambience notwithstanding, it was, after all, still the Depression:

> The Tourist Bureau, travel agencies, hotels, and other interests, publicize our Paradise of the Pacific to the end that that there shall continue to be more visitors to the Islands, to all the islands, who will enjoy their sojourn and go away to tell others what heavenly days or weeks or months they spent, sky-boating from one ocean-bound garden to another; awed by living volcanoes here, miniature Grand Canyons there; relaxing in warm velvety Waikiki waves; soothed by the murmurous gossip of scented breezes in tall and nonchalantly graceful coconut palms; meeting people from all over the world; luxuriating in the "Pink Palace" in view of surfboarding and canoing, or comfortably headquartered at hotels less ornate.
>
> But the fascinating international advertisements attract the attention of many who are not prospective tourists; which is well, if they are people prepared to transfer their business or occupation to this delightful Hawaii Territory, U.S.A. . . .
>
> It is sad, however, when men and women, boys and girls—especially girls—chuck a job in a mainland town, as soon as they have raised transportation funds (with $10 to $100 over, "to do me until I land a position").[30]

Similar to those describing jobseekers in Los Angeles some twenty years earlier, Sabin rued that "especially girls" were coming en masse. Many beat an ignominious retreat. But of course Harrison, an old hand at finding work on the road, prevailed.

Moreover, by that time arrangements were likely in place for a portion of her narrative to run in the *Atlantic Monthly*, thus promising additional income; the book would soon follow. We do not know just when and how the *Atlantic* was approached as a candidate for Harrison's work, or the negotiations that ensued. Ellery Sedgwick's papers at the Massachusetts Historical Society include no references to either Morris or Harrison, much less records of the project.

However, as early as January 9, 1931, Harrison mentioned that she and Morris had procured a "contract" of some kind (FL1), and two months later she alluded to the "publisher" who received their submission: "Their are joy in all of my letters which were sent to a Publisher in hopes that He will also for the joy that they express publish them" (FL2). "He" was undoubtedly Sedgwick, and he may have been personally known to Morris or her circle. The Morris family had left Paris for New York in the mid-1930s, facilitating communication with the Boston magazine.

Harrison followed her usual pattern by first staying at the YWCA, now in a new building on Richards Street facing Iolani Palace, the grand former residence of King Kalākaua and his successor, Queen Lili'uokalani. As its choice location shows, Honolulu took its YWCA seriously. Early sponsors were *haole* civic leaders, and they recruited Julia Morgan for its design, at the time the nation's leading female architect and now best known for Hearst Castle in San Simeon, California. A marble-floored, airy complex built around gardens and a pool, for Honolulu residents of a certain class, YWCA Laniākea offered the pinnacle of club entertainment, even as it was a haven for working-class women.

Although she soon found lodging elsewhere, Harrison returned to Laniākea for its employment agency. There, she reports, she met unaccustomed resistance: "When I went to the American Y.W.C.A. the Lady in charge of the Employment said why did you come here. I advise you to go back as the white People here want only Japanese help Well when I got through talking to her She thought very different as if any nation can keep me from getting a job and the Kind and Place and price I want" (311–12). The woman's question, "why did you come here," refers to not the agency but Hawaii itself. Overseas, being an American and a native English speaker put Harrison at an advantage. Able to fill niche positions, she was especially in demand among expatriates. In Spain, an American family paid her munificently to translate their needs. In Cairo, her lack of formal education notwithstanding, Egyptians were keen to have her teach their children. In Honolulu, however, she had to contend with different labor dynamics. For one, the territory still had an astronomical unemployment rate. Moreover, she was the wrong nationality for ready employment, and perhaps also the wrong race. In accord with the local ethnic hierarchy, in Honolulu a white male office worker made nearly double the salary of a Japanese woman who cooked and cleaned, who in turn made 11 percent more than her Filipina counterpart.[31] Neither Caucasian white-collar worker nor Asian domestic, just how did this African American world traveler, newly minted author, and "accomplished lady's maid" (ix) fit into Honolulu's labor economy?

Harrison's allusion to "any nation" who might "keep" her "from getting a job" supports two interpretations. It could refer to white Americans who preferred to hire Japanese servants. Yet it could also refer to the Japanese themselves, blocking her from due employment. Further inflecting how we read this negotiation, in light of the distancing way in which she alludes to "the white people here," her interlocutor herself may not have been white, but instead Hawaiian or Asian. One wonders just what Harrison said to her. Perhaps she made an impassioned plea for racial justice. Perhaps she gave an effective speech itemizing her skills. Or perhaps she pointed out that while Japanese women, who had a reputation for deference among employers in Hawaii, might be exemplary maids, she was not that kind of maid. More expensive, more independent, she had talents of a different order.

Indeed, she swiftly landed a "perfect job" (312) in the household of a naval lieutenant and his wife. In the 1930s, Honolulu's upper class was still very small, and only an elite minority would have employed a live-in housekeeper. As a high-ranking officer, the lieutenant had social status proximate to the islands' oligarchy (a naval lieutenant ranks considerably higher than an army lieutenant). Thus the couple would have been expected to entertain often and lavishly. Even so, Harrison had "an understanding with the Lady" (312) that she hire out the cooking for any parties larger than four. Her ability to refuse onerous duties proves her desirability, which by then may have been due as much to her celebrity as to her domestic skills, more a worldly companion or even trophy. According to Harrison, "The Lieut. wife are more than glad to have me and many of Her Friends enver Her" (312).

The lieutenant was stationed at Fort Shafter. Built on what had once been Hawaiian royal lands, the base solidified Hawaii's position as the prime US military site in the Pacific, a refueling station for ships bound for the Philippines. American military presence in Hawaii had been fortified in response to Japan's escalating colonial ambitions, shockingly realized in its invasion of Manchuria, and the armed forces stationed there swelled to 48,000.[32] Of her employer, Harrison notes, "He is on a Submarine and left yesterday for 3 weeks of war just as though they were out after enimes.... Its all because They expect war some day not to far off" (312). Indirectly funded by the US government, as so often during her wayfaring, she benefited from the imperial activities of world powers.

The military couple's home was in the Diamond Head Terrace subdivision, a sixteen-acre tract plotted in the 1920s and 1930s to accommodate an increasingly affluent white population. The subdivision is situated at the base of Diamond Head just below La Pietra, the sprawling estate owned by Walter Dillingham that was modeled on the Florentine renaissance villa of the same name. One of Waikiki's few extant historic neighborhoods, the subdivision's original Craftsman bungalows were soon joined by larger homes in a raft of styles, itemized by the Historic Hawaii Foundation as "Spanish Colonial/Spanish Mission Revival, Monterey, Mediterranean Revival, Tudor/French Norman Cottages, Colonial Revival, Hawaiian Style, and Cotswald Cottage."[33]

Given its roomy verandas and ocean vistas, my bet is they lived at 3014 Hibiscus Drive, newly built in 1935. Harrison appreciatively imparts, "This house are one of the 3 that are the highest upon the frount of Dimond Head and for that reason have 5 diffrent Terraces on 3 sides all with a delightful View" (312). She reveled in its location, "right in the Playgrowns of Waikiki in 3 blocks of the Sea and two of the Picnice Growns Polo Growns Tennis Court & Band Park" (313). It was "as much like Cap d'Antibes if it had been made to order it was the one thing I was more anxious to get so that I wouldn't feel lonely for the beautiful soroundings that I had left in France" (312), she states. Her nostalgia assuaged, her ardor for a new idyll sparked, she iterates, "so I didnt need to remain at Cap D'Antibes to have a happy Home" (316).

Harrison concludes her account of the position by quoting her employer: "She said the important part is that I take care of Them so beautiful and the House" (312). Yet despite this seeming endorsement of providing "beautiful" care, she quit as soon as she could afford to. On the strength of her new authorship, she left that commodious household to embark on a very different lifestyle in a very different kind of home. The epilogue opens with the disclosure, "That cheque from the Atlantic Monthly for my article gave joy. I got it on a Sat. and gave up my weekly job. This is what I am during with the money as I want to enjoy every penny. I had a Tent made to order of course This was my plannes before I left Cap D' Antibes so they have been carried out perfect" (315). Some of the *Atlantic* money also went toward "a few Hula lessons" and a "serfe board," "to add gayness to that list of things the check bought" (318).

Finding the right site for her custom-made "7 × 7 orange Tent" (317) was paramount. She recounts, "I went around Waikiki looking for a privat Yard to Put my Tent So that I could be in a privat Place yet free good and cheap. I asked several just to see no one Knew of any place but I was only Testing that alway give encouragement. I choosed the best from several right here at Waikiki in the front Lot of a nice Japanese Family" (315). That nice family was Masakichi and Kuni Tada and their five children, soon to be six. We can wonder about the couple's response to what must have been a surprising request, and the private discussions they had as they weighed the prospect of Harrison joining them at 105 Ohua Avenue.[34] Inventorying its advantages, she continues, "They have a Cottage in the rear and the Show bath and W.C. are in the Yard and I am right at the front gate so do not trouble any one freedom is my main point. My grown for my Tent cost $2.50 a month" (314).

Facing "the Beautiful perfect Hawaiain St. Augustine Church," her new home was at the far edge of Waikiki, a scant two blocks from the beach. She declares it to be "the best neighbor hood" (314). The church has since been rebuilt, while the site of Harrison's tent and the Tadas' bungalow has undergone a much greater, and rather eerie, transformation—subsumed by the Marriot Beach Waikiki Resort and Spa, which both sprawls and looms. Already by the 1930s, modest private properties like the Tadas' had begun to give way to corporate development, and the quarter is now oriented toward tourists. Nevertheless, it retains a more casual and residential feel than other parts of Waikiki. The beach is still a local favorite; the ocean is unchanged.

Harrison lived there for four and a half years, up until she left Hawaii in late December 1939. The youngest Tada was born shortly after she moved in, and over the course of her residence, the oldest began and completed high school. She must have come to know that family of eight quite well. She was not just proximate to them, but to a degree lived with them, and not only because the shower and toilet were communal. Given Harrison's usual restlessness, her long stay testifies to a very good relationship. Even as her most recent employers were bracing for war against Japan, and even as she was aggravated by competition from local Japanese workers, she relished life with a Japanese family. Her Christmas habit was to tote a pillowcase downtown to the local drugstore, to fill with "toys and candies" for "the children in the Yard."[35]

In the epilogue, Harrison describes the tent and her unorthodox, semi-outdoor housekeeping in close detail:

> It have Two windows at one end and one at the other and one pocket to stick my flouer vase it open at each end the same way like what are used on sport shirts instead of laped over or laced up and on Each wall which are 4 feet are a pocket for my cloths and a strip over the pocket for to hang a pretty bright Curtain. I found a boy that made me a floor for a dollar but use His Father's lumber. he lives on my street. the floor is good but light so I can carry it on my head when I want to move. I have a green linolien on the floor with two small rugs I have a good army cot and a new matrews made to order. a beautiful bright Chinese blanket a good pilla bright ticken and two sheets and one pilla case made of outen flannel bright print. (317)

She adds, "When the light is out and the door and Windows closed the lights of the street shine through the holes and on to the Top of my Tent and it look just like the Stars" (318). "Furnished nothing like a home," "look[ing] like a poch" (317), the tent is both porous and portable. "I can take up my house and walk at any time," she enthuses.

The symbolic resonance of the tent feels inexhaustible. Purchased with Harrison's first writing income, it is a material emblem of her authorship. It at once marks the end of her world journey and promises future movement. It shows her commitment to breaking down hard divisions between inside and outside and between public and private. Its liminal quality is mirrored in the larger liminal space of the nearby beach she haunted—where land and ocean meet—as well as, scaling out, is mirrored in the politically shifting space of a militarily strategic Pacific island that was American territory but not yet a state, home to a majority immigrant population that would soon have to prove its patriotism.

Harrison named the tent "Villa Petit Peep," hearkening back to her agreeable stint of housesitting at "Villa des Verveines" on the French Riviera. The invocation of a foreign domestic model, "villa," is perhaps another sign of how her own country had failed her. Equally significant is the actual name, "Petit Peep." Both gazed at and gazing, she was registered as a tourist attraction even as she participated in tourist activities herself. Her explanation for the choice holds pride of place in her book as its final lines: "This is my first and only Home. Villa Petit Peep are The name of my Tent as I let my callers sit on a seat in the yard and Peep in so I gave it this True name" (318). Closing the volume, the statement can be read as commentary on not only her home but also her authorship. With the publication of *My Great, Wide, Beautiful World*, she opened up her life even further, to the scrutiny of the public at large.

Her choice location enabled Harrison to experience Hawaiian culture in far less commercial form than the "real Hawaiian Picnice" she attended on first arrival. "Every Friday evening," she states, "I go to one of the high class Hula Schools to see the student dance its free" (313). Allowing for even more regular, and expert, exposure, she lived right next to the Lalani Hawaiian Village, a family enterprise founded in 1932 by master ukulele maker and entrepreneur George Mossman. She could see and hear the village's performances from her tent.

While by 1955 the concept of the "Hawaiian Village" had gone corporate with the completion of the Kaiser Hawaii Village Hotel (now Hilton Hawaiian Village), these facsimile settlements started out as local ventures managed by families and small community groups. Many were ephemeral, with shifting locations and participants. Lalani Hawaiian Village was the largest and best known. Designed to replicate a traditional Hawaiian community prior to Western contact, it was composed of a cluster of buildings erected on a one-acre lot on the corner of Kalākaua and Paoakalani Avenues. Adria L. Imada maps it thus: "Built by a master Hawaiian hut-maker, the village included seven hale pili (grass huts) for sleeping, cooking, eating, worshipping, kapa (bark cloth) weaving, and storing wa'a (canoes). Tall coconut palm trees grew between the huts and a large kii (wooden carving of a god) stood in one corner."[36] Visitors watched dances and heard live music, listened to talks, and ate Hawaiian food; the more committed could receive instruction in the Hawaiian language, music, hula, cooking, surfing, or fishing.

Serving up ethnographic entertainment to paying guests, the village shared some traits with Harrison's beloved world's fairs. Yet as Imada points out, as a "self-supporting enterprise" it also implemented the "small-scale family-based agricultural practices that preceded the incorporation of large tracts of land into industrial sugarcane and pineapple plantations."[37] The Mossmans' aim, moreover, was to safeguard Indigenous culture even as they made a living. By 1934, the village was home to eighty-seven-year-old James Kapihe Palea Kuluwaimaka, the former royal court chanter who could recite genealogies of over sixteen hundred lines. Astonishingly, this direct link to Hawaii's royal past was Harrison's neighbor.[38] In turn, Hawaii's future was augured by Mossman's daughter Pualani, a budding star who was soon to become "the most photographed girl in the islands" in her capacity as actor and model for, respectively, the Hawaii Tourist Bureau and Matson Navigation. Of the family business, Pualani recalled, "People from all the hotels would come down so happy to see our beautiful show, the imu, taste real Hawaiian food, see the boys climbing the coconut trees, perform the knife and fire dances. All for $1 or $2."[39]

For Harrison, it cost nothing at all. Underscoring once again her ability to extract gratis entertainment, she states, "Just back of me are the Beautiful Hawaiain Village so I get plenty Hula musice and dancing free" (316). In the very same sentence, she continues, "when there is a Ball in the Garden of the Royal Hawaiain Hotel facing the Sea we swim or wade and look right on the dancers and all the other things of Pleasure right near where I set up my Tent" (316). Her juxtaposition posits the Hawaiian performers at Lalani Hawaiian Village and the upper-class guests at the Royal Hawaiian Hotel as equivalent spectacle, each providing a culturally specific form of dance for her delectation.

The exclusive hotel was the image of Waikiki glamor. "Highlights on Honolulu," a 1930 promotional pamphlet, urged, "If you are going to be very grand, go to the Royal Hawaiian Hotel. It's quite the last word."[40] Its guests included the likes of Nelson Rockefeller, Babe Ruth, Doris Duke, Shirley Temple, Douglas Fairbanks, and Mary Pickford—whose film *Sparrows* Harrison had enjoyed in Dijon. For the islands' wealthy residents, the Saturday night dance was a weekly highlight.[41] Just who composed the "we" Harrison stood among as she gazed at the dancers from the shallows, the fine white sand beneath her feet—curious locals, eager tourists, other adventurers like herself?

As so evident here, Harrison liked to be around affluence. She materially benefited from rich people's surplus, their hand-me-down clothes and their vacant homes. But such bounty aside, she registered the rich as an aesthetic spectacle in and of themselves, taking pleasure in their good looks, plush appointments, and lavish entertainments. As a way of accounting for her contentment in Waikiki, she states, "I want alway to be where wealth health youth beauty and gayness are altho I need very little for myself I just want to be in the midst of it" (318). The first item in her list is "wealth." Far from denouncing systemic inequality, she only criticized the stingy habits of those who could afford to spend well.

While relaxing at 'Kapi'olani Regional Park—just a stroll from her tent— Harrison penned a voluble letter to Alice Foster about her island life. The park's lush seaside expanse at the base of Diamond Head Crater, once the site of a racetrack founded by King Kalākaua and later an enclave for Honolulu's wealthiest residents, opened to the public in 1896. We can envision her on its vibrant green grounds, writing under a banyan tree by the tennis courts while the Honolulu Cricket Club went about their practice.

Once again showing Harrison's proximity to elite practice and imagination, that same park has a cameo in *The Great Gatsby*. Referring to their Waikiki honeymoon, Tom angrily challenges Daisy's denial of having ever loved him. "Not at Kapiolani?" he demands. "Not that day I carried you down from the Punch Bowl to keep your shoes dry?"[42] In the novel he wrote on the French Riviera not long before Harrison settled there, Fitzgerald imagined her future home as a romantic destination for his fabulously rich newlyweds. Yet her daily life in Waikiki rivaled any honeymoon. Her regular 6 a.m. ocean swim was preceded by a breakfast of pineapple and followed by a "bath right there under the Show on the Beach" (316). She habitually cooked and slept outside and acquired a dog, Pluto, who accompanied her about town (to judge by its white coat and black markings, a spaniel or terrier mix). With royalties from her book streaming in, for a time, at least, the rhythm of her days was determined by not the clock but the sun, her time to command as she pleased.

Typical of Harrison, her satisfaction depended on not only Hawaii's actual topography but also its cultural depth, especially as formed by its Asian populations. Although a remark about the construction of her tent "made just as I wanted it" signals the islands' racial hierarchies—"A fat German where I got it was so Kind and let the Chinese Girls that Sew for him do all the little extra work" (317)—Harrison was more likely to socialize with Asian people than to be served by them. She comments on her enthusiasm for "the Change of going to Chinese gatherns, Japanese" (314), which she saw as fair recompense for the homey cafés she had frequented in France. The *Honolulu Star-Bulletin* verifies her connection to the city's Chinese community in recording that in October 1937 she attended the eighty-third birthday party of one Mau Wong, held at the Wongs' Duval Street residence. The majority of the fifty-plus guests were Chinese, and Harrison was one of only two who did not have a Chinese, Hawaiian, or Portuguese surname.[43] Her attendance at such a party confirms both her popularity and her singularity.

Living in Hawaii also strengthened her ties to Japan, a country for which she had special feelings. "I have been longing to reach just this spot" (286) she had exclaimed from Kobe in the fall of 1934, scarcely half a year before settling in Honolulu. Her book chronicles an enjoyable two months touring Honshu, with a period of residence in Kobe and stops in Kyoto, Osaka, Nara, Kanagawa, and Tokyo. Perhaps it was in Japan that she finally realized the "bright vision of templed cities" (ix) she had had as a child worker in Mississippi, awakening travel dreams. However, she only took a day trip to the nation's premier templed city, Kyoto. Rather than pass additional time in the ancient capital, she lingered forty-five miles to the west in Kobe, the port city still known for its easygoing atmosphere and salient internationalism. In Kobe she enjoyed cherry blossoms, forest walks, kabuki, and the gregarious public baths, along with the views from her room at the YWCA, which saw the mountains framed in her front window and the sea, in the bay one. She found ready acceptance among the Japanese, a curiosity who could "amuse" even as she was "being amused" (286), as she put it. Subsequently on visiting Taiwan, occupied by Japan since 1895, she had the opportunity to experience its culture exported to another land, a mode that always appealed to her.

In Hawaii she could likewise live among Japanese people outside Japan. At the time, the Japanese comprised approximately 37 percent of Oahu's population. The Meiji Restoration of 1868 had abolished the Tokugawa shogunate and its feudal order, ushering in Japan's modern era, along with profound social disruption. Displaced samurai took to the road, and countless farmers were made landless. Hawaii's sugar industry, always hungry for workers, welcomed this labor surplus, even as the Japanese government sought to temper the crisis by funding mass emigration. Scott Kurashige identifies the citizens who emigrated during this tumultuous era as a "fallen middle class."[44] The majority were from the Okinawan island chain, the island of Kyushu, and the prefectures of Okayama, Hiroshima, and Yamaguchi in western Honshu. Harrison's hosts, Masakichi and Kuni Tada, hailed from Yamaguchi, still one of Japan's most rural prefectures.

According to the "List of Manifest of Alien Passengers," on August 30, 1918, Masakichi left Yokohama for Honolulu, where he joined his younger brother. This was in fact a homecoming, as he had already lived in Hawaii for a good twenty years, initially settling in Maui in 1896.[45] He had returned to Yamaguchi in 1916 scant weeks after his first wife, Taki, died. In Yamaguchi he married Kuni, twenty-two years younger than he and from the same small township, before he set sail again for Honolulu. The newlyweds were separated for several years, as Kuni only made the Pacific voyage in May 1921, her passage paid for by her husband.[46] Fortunately she arrived prior to the Immigration Act of 1924, which consigned many first-generation Japanese men to virtual bachelorhood. Women of their own race were prohibited from entering US territory, even as custom and sometimes law (albeit not in Hawaii) checked them from marrying women from other ones.

By October 1918, Masakichi had begun his long-term service career with businessman James Steiner. He worked as a cook and "butler" at a "private home," and Kuni supplemented their income by taking in laundry.[47] Their large family, four girls and two boys, was typical for working-class Japanese. Harrison's esteem for Japanese childrearing underpinned her relationship with Kuni. "Of all nations the Japanese women are only worth being called Mother" (291), she had declared in her book. "I wouldnt give One Japanese mother for the American ones." For their part, the Tadas must have found it cheering that their lodger had spent a stretch of time in their native country, an experience that would distinguish any American in the 1930s, much less a working-class woman. Moreover, in sailing from Japan to Hawaii, Harrison had made a voyage similar to theirs, and she went through the same process of arriving in a new place, settling in, and finding work, although in her case as a citizen rather than immigrant.

The history of Tada's employer reads as a tale of immigrant triumph and, more specifically, of *haole* capitalist success. Steiner was a migrant from Czechoslovakia who came to Honolulu by way of Missouri, working in restaurants. He went on to open the territory's first ice cream parlor, becoming, at least in his own assessment, its "Ice Cream and Candy King." Over time, the enterprise became the Island Curio Company, said to purvey "native curios from almost every part of the island." Steiner specialized in postcards, and one of the few artifacts inscribed in Harrison's hand is a postcard of the Moana Hotel published by his company.[48] Perhaps Tada had passed it on to her.

Steiner also bought up Waikiki beachfront real estate, at a time when it was seen as too remote to be commercially viable. The mansion he had built there, and in which Tada worked for decades, broadcast his achievement. "Colonial in style, modified to suit the tropical climate" and featuring "many innovations in the building craft," the estate was built on land that he purchased by auction from the US territorial government.[49] It became the Halekai Officers Club during World War II.

Confirming his status as family retainer, Masakichi was the only one of Steiner's servants to receive a legacy on his death in 1939.[50] His $250 inheritance would be worth about $5000 today, not especially munificent—Steiner's four sons each got $20,000—but not insignificant either. Comparing the rise of these two immigrant men is instructive. Tada had a stable job, and he and his family resided in a pleasant bungalow not far from Steiner's home. By all indications, Tada's children entered the middle class, and his grandchildren indisputably did. Still, differences in nationality and race ensured that his rise was modest in comparison to that of his employer and other *haole* men who similarly ascended from humble beginnings to great influence in Hawaii.

Harrison's trajectory in Honolulu of course resembled neither man's. Even as authorship transformed her social identity, in many respects this "vagabond housemaid" lived much the same way she always had. The next chapter discusses the reception of her book, and the effect it had on her.

CHAPTER 8

Authorship and Fame

Harrison writes proudly about the success of "my Book" in a letter to one Mr. Clay in Vancouver, British Columbia. Headed, "Tuesday May 17, night 11.15 p.m.—1937 Sitting on the floor of my tent," in full it reads,

> Thank you Mr. Clay, for the Four Partridges i received them in time for Christmas i think. i went to bed early and as i need so little sleep i am up. But will be ready for another sleep about 4 a.m. i am writing to say be on the look out for Mr. and Mrs. Tragella. Their home are in Canada but He is the Head waiter at the Royal Hawaiian and Madame are at the Sister Hotel Moana Seaside in one block along the beach of each other she make the reservations there. But they are more than that they are the Kindest Two and try always to help those that need it and help them in a good and grand way. You'll enjoy looking at them. She have my Book it have sold wonderful but i am still a maid not every day but as often as i need a few cents for my daily bread. I must keep the book earning for when i am in my 30's cannot be 20 all the time.[1]

The letter testifies to Harrison's international circuits of patronage and contiguity to elite circles. Befriended and befriending, she was on familiar terms with women and men with rarefied lifestyles, people who in small ways sought to enrich hers: Mr. Clay, keen bird hunter, was affluent enough to ship some of his trophies all the way to Honolulu. We see too her connections with Oahu professionals, on congenial enough footing that she would promote their careers; the note was to serve as the Italian couple's letter of introduction to Clay. Originally from Milan and a recent arrival to Honolulu by way of Vancouver, Umberto Tragella was one of the many European artisans recruited by the islands' fledgling resort industry, especially Italian and German chefs, architects, and masons. Starting out as the maître d'hôtel at the Royal Hawaiian Hotel, Tragella would become an influential purchasing agent for Matson.[2]

The same letter shows the role that authorship played in Harrison's finances, along with her continued need to take domestic jobs: that somewhat rueful confession, "but i am still a maid." Even at the height of its sales, the book permitted her only to cut back on her hours, and she suggests that she may have hired out her services by the day. Testament to her dual professions, the 1937 Honolulu city directory gives her occupation as that of maid, while the following year it has her as a writer. Just what was the nature of her book contract? Were the terms she agreed to fair? If the book "sold wonderful," one would expect the royalties to be more than enough to support her frugal lifestyle. Her rent, after all, was only $2.50 a month.

Harrison's alacrity to quit jobs as soon as she could afford to, her hope to scrape by in Honolulu with "a half time place" (311), and her anxious joke about a future in which she relied on such work—"I must keep the book earning for when i am in my 30's cannot be 20 all the time"—show how onerous she could find domestic service.

Nor though did she wholeheartedly embrace a writing career. Her enthusiasm for authorship and a public profile waxed and waned. Long before her book came out, Harrison had declared of her writing, "Like my traveling I did not do it for the Public it was for my own pleasure and I do not care weather the Public know who I am" (FL2). At first she had even planned to publish under a pseudonym. Nevertheless, she happily received the fans who sought her out in Waikiki, and she actively, if sporadically, fanned her publicity. While a Los Angeles journalist described her as "haunted by nosey autograph hunters,"[3] she herself approvingly noted, "I have autographed for many fine People Visiting here" (FL3). Authorship ushered in a new kind of correspondence, writing to strangers. "I enjoy it and nothing else of the book," she claimed, and she indicates that her satisfaction had an element of national pride in stating, "Since the book came out I have rec'd letters from dozon of fine unseen friends of Americans best and they are so true at heart from the reading of it" (FL3). She also remarked, "I wish that the Book could be sold for much less I like best to give pleasure I need so little change."

The confiding holiday letter she sent to Ellery Sedgwick at the close of 1936 is even more revealing than the epistle to Clay. *Atlantic Monthly* readers got one last view of Harrison in Honolulu when it ran in the magazine's March

1937 issue. Long accustomed to associating with "the finest people," to use her phrase, she wrote to the acclaimed editor with considerable familiarity and ease—evidently, not for the first time. In full, the letter reads:

> Dear Mr. Sedgwick
>
> may this find all well and enjoying the new year, i am in the same spot as this time last year and just as soon as i finish this will take the same pillow case to our beloved Kress for toys and candies for the children here in the Yard. it is more private now a fence around with a gate and my tent are souronded with green plants and i like my Tent life more than ever i worked up at Black Point Road 3 months and only slept well when i came down to my Tent, did You like the way the Book are made up? Mrs Dickinson liked it which pleased me more than any thing. I didn't want it sold here i like best to be about and see but not be seen and the finest people have called to see me. it have sold well here and would have sold many more this month but the last was sold Thanksgiven day and the order didn't get out the strike have made it very bad for the Islands but it is very Christmases and quite cool, i receive so many delightful letters from readers of M G W B W and laugh when they ask how soon will i have another one. i just side step questions like that i dont know how soon we will get some change maybe next may. Miss Mildred borrowed it from her Mama and are dissipointed to wait so long, i am so busy enjoying my back to nature life i wont think of it until my Tent begain to leak, i hope that You have one of my Great Wide beautiful World. Yours truly
>
> JUANITA[4]

Expressing her continued satisfaction with her "back to nature life," Harrison stresses her warm relationship with the "children here in the Yard" and claims an insider Honolulu identity through referring to "our beloved" five-and-dime. She also reveals her mixed feelings about her publicity. On the one hand, she would prefer to walk about unrecognized. Yet on the other hand, she was gratified by the visits and accolades and found her fan mail "delightful." She felt no urgency to "have another" book; only when she needed more money to live on would she turn her hand to it. She had recently worked in the ritzy Black Point

Road neighborhood, a cliff-top cluster of houses overlooking an exceptionally turbulent surf that was soon to boast a mansion built for Doris Day, but her book sales were robust enough—the nationwide seamen's strike of fall 1937 notwithstanding—for her to quit that job and live, for a time, off her royalties.

We learn a cozy detail, that she referred to her book as "M G W B W." Her interest in Sedgwick's opinion about "the way the Book are made up" prompts many unanswered questions. Finally, the missive confirms her ongoing communication with both "Miss Mildred" and "Mrs Dickinson" and forecasts her summer 1937 reunion with the latter.

The *Atlantic Monthly* was so prestigious that on the strength of its two excerpts, Honolulu newspapers had run profiles on Harrison even before *My Great, Wide, Beautiful World* came out. Once it did, the *Atlantic* preview and Macmillan's aggressive promotion made for brisk sales. "It were a best seller" (FL3), Harrison boasted. Tallies published by the *New York Herald Tribune* and *Los Angeles Times* back her up. While the term "bestseller" is notoriously imprecise, it does at least indicate a "better seller." *My Great, Wide, Beautiful World* went through a rapid succession of printings: nine between May 1936 and February 1937 (including a Canada edition), and then a final one in February 1939. An excerpt comprised the Book Supplement for the February 1937 issue of *Reader's Digest*—she had really made it. And yet: "But i am still a maid."

Harrison's is among a fleet of travel books, both memoirs and guides, published in the United States in the mid-thirties. Examples include Harry A. Franck's *A Vagabond in Sovietland* and *Roaming in Hawaii*; Graham Greene's *Journey Without Maps*; inaugural editions of Eugene Fodor's *On the Continent* and Victor Hugo Green's *The Negro Motorist Green-Book*; and Eric P. Quain's *Touring South America*, a volume of letters that was advertised alongside the "penetrating diary of an extraordinary colored woman"—that is, Harrison's book.[5] Resembling her volume for the novelty of their authorship, the Abbe children's *Around the World in Eleven Years* and the putatively as-told-to memoir *Sister of the Road: The Autobiography of Box-Car Bertha* came on its heels.

In respect to its author's race, however, *My Great, Wide, Beautiful World* was unique. Writing for the Associated Negro Press, Frank Marshall Davis assessed it as "one of the few travel books by a writer of color." Korey Garibaldi attributes its enormous commercial success—at the time unprecedented for a Black-authored book, of any genre—to the comprehensive marketing machinery put into motion on its behalf.[6] Harrison's sales eclipsed those of not only literary contemporaries such as Jean Toomer and Nella Larsen but also more popular authors such as Richard Wright. It was Harrison, not Wright, Garibaldi states, who was the first African American writer to break out of the "literary ghetto."[7]

The fierce working-class protagonist of Zora Neale Hurston's 1937 novel *Their Eyes Were Watching God* might recollect Harrison. Yet in high contrast to Hurston's and Harrison's portraits of questing, rebellious African American women, the American cultural landscape was saturated with depictions of Black subordinates. The most pernicious was Margaret Mitchell's *Gone With the Wind*, soon adapted into a Hollywood blockbuster. Appearing the exact same year as Harrison's book and advertised alongside it, the novel depicts devoted Black service taken to fantastic heights, especially through the character of Mammy.[8] William Faulkner's *Absalom, Absalom!* tracks the fortunes of the slaveholding Sutpens in Mississippi and their efforts to excise the Black members of their family. 1936 also saw the reissue of the film *Imitation of Life*, an adaptation of Fannie Hurst's novel of the same name, which explores the relationship between a white woman and her Black maid, Delilah Johnson. Together the two women run a lucrative pie business, but Johnson rejects the offer of full partnership to remain a maid. Harrison's book offered a potent antidote to these stereotyped scripts.

Her story of course did not reach readers in unmediated form. Rather, it was sculpted by a series of editorial and marketing decisions. To use John Kevin Young's formulation, Harrison's individual "zone of composition" was embedded in the "hybrid sphere of publication."[9] Her reception was directed by the paratextual apparatus that connects *My Great, Wide, Beautiful World* to its readership, as comprised by its title, frontispiece, preface, epigraph, and dust jacket design, as well as, more fundamentally, by how the letters were edited and organized. Garibaldi contends that Harrison's "racial heritage is revealed by design" through the preservation of her faulty spelling and grammar, evident as early as the dedication.[10] Moreover, recollecting the authenticating materials used to frame slave narratives, Mildred Morris is accorded a prominent position as having "Arranged and Prefaced" the manuscript. While it is absolutely true that she had done so, the phrase disposed skeptical readers to accept that Harrison really wrote the book. The *Atlantic Monthly* had gone further in openly assuring its readership, "Miss Harrison's diary is entirely genuine."[11] Judith Madera reminds us that "as much as archival histories are sites of contest, so too are individual texts," and this contestation can be racially motivated.[12]

The title itself is more multivalent. With its global claim, it positions Harrison within a long tradition of imperial around-the-world narratives. Yet the title also situates her in a distinctly American, middlebrow context, as it derives from the 1899 William Brighty Rands poem of almost the same name, "My Great, Wide, Beautiful, Wonderful World." (Apparently "Wonderful" was deemed one adjective too many.) Two verses from the same poem compose the book's epigraph. Featured in composition textbooks for older children and assigned to younger ones to memorize, it was once a grade school staple, so familiar that the heroine of *Rebecca of Sunnybrook Farm* stands up in a moving wagon to declaim its opening lines: "Great, wide, beautiful, wonderful World/ With the wonderful water round you curled/ And the wonderful grass upon your breast,/ World, you are beautifully drest!"[13] The summoning of the popular poem wraps Harrison's book in a reassuring mantle of middle-class associations, even as the nonstandard prose of the book's contents testifies to her exclusion from formal education. As a bid to narrow the gap between this Black working-class author and a presumed white middle-class readership, it is akin to Morris's emphasis on Harrison's girlishness and not recognizably African looks, her "slight form, fresh olive complexion, long hair braided about her head" (xi).

The dust jacket's photo collage of Harrison in a medley of poses, costumes, and settings—the original visual point of entry to the text—makes bolder claims. The photographs include her dressed in a traditional Palestine outfit, veiled and holding a clay pot, posed with a guitar, strolling baguette in arm on Promenade des Anglais, and standing in a swimsuit on a Waikiki beach. We also see a number of cartoon-like drawings of men and women and patches of illegible writing in her hand, the meaning hovering tantalizingly just out of reach.

This deliberately modern presentation does much to dissolve the residue of gentility that Rands's verses spackle onto the book. It offers, moreover, further evidence of how Harrison's racially indeterminate appearance and penchant for costumes supported her fluid social movement. What a difference between these deliberately chosen outfits and the "woman's cast-off apparel" (ix) that was her lot as a child in Mississippi! By posing thus, Harrison certainly participated in questionable Orientalist practice. The assembly of props argues for mainstream status: That she dresses up as an exotic proves she isn't one. At the same time, however, her avid costuming is still another form of self-determination. In both her text and her lived practice, she claimed a ruling-class prerogative: the "right to be various," as Richard Dyer formulates it in his landmark study *White*.[14] The cover gives visible form to a personal geography of perpetual interplay between different places, cultures, and ethnicities and to the multifaceted subjectivity this geography produced.

Harrison once proudly stated she had been "in nearly every state in the Union." For years she had "traveled about from State to State City to City" (FL2),

FIGURE 4. *My Great, Wide, Beautiful World*, original cover. Source: Courtesy Between the Covers Rare Books.

spending time in all the nation's regions, and now her book was read, reviewed, and discussed in the same. A sampling of newspapers in which reviews of it appear includes *The Brooklyn Daily Eagle*, *Central New Jersey Home News*, *Nashville Banner*, *El Paso Herald-Post*, *The Morning Star* in Manhattan, Kansas, *The Province* in Vancouver, British Columbia, and *The Napa Journal* in Napa, California; there are countless more.

Even a passing survey of reviews proves the diversity of needs *My Great, Wide, Beautiful World* met for different readerships and how variously it was situated. Every last one mentions its author's race. Some of the language is egregious, as when Harrison is identified as a "dusky lady."[15] Yet often her racial identity is presented as an essential biographical fact that while noted does not control the evaluation. *Smart Set* editor Charles Hanson Towne wrote a standard, and yet quite affirmative, review that opens, "Not often will you come across a travel book the equal of Juanita Harrison's 'My Great Wide Beautiful World.'" Drawing liberally from Morris's preface, he continues, "Here is an uneducated colored woman of 36, who worked at almost any domestic job she could find, and always had a secret longing to see strange lands.... She fell in, fortunately with some good people in California and began, under expert advice, to invest her small capital, until she found herself in possession of a meager $200 a year. Undaunted, she packed her suitcase in 1927 and so began her wanderings in many lands. She kept a diary, which has been edited by Mildred Morris, but not, obviously, greatly altered, and it makes beautiful reading. For Miss Harrison is that rare thing, a natural writer."[16] The review is typical in that it leads with her race, emphasizes the offices of the "good people in California"—the Dickinsons—and opines that Harrison is a "natural writer." Nevertheless, Towne, himself a successful poet, is emphatic in his admiration of the book's literary quality, "beautiful reading." In Honolulu his review was reproduced—and thereby framed in new ways—in the Japanese American newspaper *Nippu Jiji*, which as we will see took a proprietary interest in Harrison.

Carl Van Vechten deemed this innovative, expatriate African American life writer to be thematically and stylistically fit to introduce to his modernist friends, enjoining both Gertrude Stein and Langston Hughes to read her *Atlantic Monthly* excerpts. Hughes went on to urge her book on his patron Noel Sullivan. "Have you read My Great Wide Beautiful World by Juanita Harrison, the colored woman who worked her way around the world?" he asked him. "By all means get it, as everyone says it is delightful. I bought it, but haven't read more than a few pages yet, but they were swell." Just who was that "everyone" who found the book "delightful"? Other writers like themselves? Parenthetically, he continues, "(I would have sent it to you along with the Mann stories, but since it's been out for several weeks, I was afraid maybe you had it.) But if you haven't got it, do get it soon. It might be something nice for Marie to read, or does she feel well enough to read? I will send it to her if you think she would like it . . . . . . [sic]."[17] Hughes both couples Harrison's narrative with high literary fare—the other book to which he refers is Thomas Mann's *Stories of Three Decades*—and posits it as therapeutic for the unwell Marie. Indicating the special interest *My Great, Wide, Beautiful World* had for those who followed African American literature, of note too is his belief that since "weeks" had passed since its release, Sullivan may well have already had it.

No wonder Hughes admired Harrison, given not only their shared restlessness but also their shared fearlessness about challenging notions of race responsibility. In his manifesto, "The Negro Artist and the Racial Mountain," Hughes famously declares, "The Negro artist works against an undertow of sharp criticism and misunderstanding from his own group and unintentional bribes from the whites. 'O, be respectable, write about nice people, show how good we are,' say the Negroes. 'Be stereotyped, don't go too far, don't shatter our illusions about you, don't amuse us too seriously. We will pay you,' say the whites." Urging resistance, he resolves, "If white people are pleased we are glad. If they are not, it doesn't matter. . . . If coloured people are pleased we are glad. If they are not, their displeasure doesn't matter either."[18] In accord with such a perspective, while Harrison certainly sought to appease the individuals to whom she wrote, nothing indicates that she viewed herself as a representative African American woman or felt the constraints that such an authorial identity could impose.

Some of the reactions to her foregrounding of sharp appetites and shrewd ways recollect the "displeasure" Hughes invokes. The sharpest is Alain Locke's: "As to *My Great Wide Beautiful World*, the significance fortunately is only individual: an illiterate carried round the world is at the end of the trip, and in the volume that reports it, is illiterate still."[19] Probably Locke would not have been pleased that in the letter to Sullivan, Hughes mentions him and Harrison in almost the same breath. More sympathetically, Arthur B. Spingarn, the white civil rights activist, voiced concern that her book be marshaled to support racial stereotypes, allowing, "A radiant and valiant personality shines through the book, but though highly touted, it is likely to thrill only those people who know 'colored people are like that.'"[20]

White women's book clubs gave Harrison a surprising amount of airtime, in ways that might confirm Spingarn's rueful prediction. She was discussed in club meetings across the land, from California to Minnesota to Florida. At a Tampa meeting, one Peggy Miller "entertained with excerpts" from her book, while the *St. Cloud Times* reported on "a most entertaining review" presented by a club member with the help of a map.[21] The consistent emphasis on entertainment raises the question of whether these women indulged in racial minstrelsy. Did they adopt exaggerated accents as they read aloud from *My Great, Wide, Beautiful World*? When white readers responded to Harrison, they trod a perilous line— approached and sometimes crossed—between appreciation and condescension, or worse.

William Pickens's record of his voyage to South America, previously discussed, displays a quite different perspective in identifying Harrison's book as absorbing fare that along with the sublime seascape distracted him from his own writing. "If I can stop long enough from contemplating the sea and the clouds and the stars and the 'Great Wide Beautiful World,'" he confides, "I shall try to tell you something about some of those places."[22] Pickens pairs her book with his other shipboard reading, "Heavenly Discourse," the daring philosophical dialogues of Charles Erskine Scott Wood. Like Hughes, he shows that *My Great, Wide, Beautiful World* could find a place on the programs of quite serious readers.

That Harrison was living in Hawaii just as the book came out fanned her renown. Had she settled elsewhere, she would have attracted far less attention. Her location also expedited personal encounters with admirers, since Oahu was both an upper-class holiday destination and a Pacific transit hub. Pickens had actually visited Harrison in Waikiki just prior to that trip to South America, an occasion covered by the *California Eagle*. The article alludes to the meals that "the author and a friend" prepared "in the open ovens along the beach," the activist's discomfort about changing his clothes outside, and his host's predilection for walking barefoot through town.[23] Gwen Dew, special correspondent for the United Press—and soon to be prisoner of war in Japan—also met with Harrison. Another visit that made the news was that of Carl Brandien, an itinerant painter notorious for toting all his worldly goods in a backpack. (Given that it weighed in at fifty pounds, Harrison was surely not impressed.) In Waikiki, Brandien painted impressionistic, romantic beach scenes featuring sea and sky, and he painted Harrison, too. A *Honolulu Star-Bulletin* account of their meeting enthused, "There was just one person Carl W. Brandien, vagabond artist, wanted to meet when he arrived here six weeks ago. That was Juanita Harrison, vagabond author of *My Great Wide Beautiful World*."[24]

As a celebrated local transplant who was both author and maid, Harrison occupied a unique and rather strange position in Honolulu. She was a source of pride for some Honolulu residents, gratified that an "Authoress Living Here" was "Praised by Mainland Reviewers."[25] Enid Northwood, a novelist and playwright based in Hawaii, maintained, "Juanita Harrison, colorful housemaid who has now settled down permanently in Honolulu, can almost be claimed by the Islands."[26] The *Honolulu Advertiser* singled her out as a "striking example of the interesting personalities that come here," and she is listed as one of the sights of Lei Day, a newly established event meant to commemorate traditional Hawaiian culture: "Mrs. Charles F. Chillingworth sniffing her fragrant crown lei ... Officer Van Poole carrying off the May Pole ... Juanita Harrison, vagabond housemaid on her way around the world, asleep on the front lawn of Honolulu Hale."[27] Note the suggestion that Honolulu was just a resting point in her travels, opposing the claim that she was there "permanently."

A *Honolulu Star-Bulletin* profile by William Norwood presents her expansive domestic life as seamlessly joining her creative endeavors, contending, "Juanita spends most of her time on the beach, where she has her kitchen and where she communes with her various muses. There's a place along the seawall where the waves have formed a cave, and there Juanita prepares her vegetables."[28] Coupled with her travels, such habits made her a point of reference for hardy women iconoclasts in the Hawaii hinterland. A farmer in Holualoa who turned to *My Great, Wide, Beautiful World* for bedtime reading and journal inspiration confessed to wishing she could display as much "good humor" as Harrison and have as much "exciting" activity to write about.[29] Proving her enduring imprint on island memory, as late as 1966 she was invoked to frame the rugged experience of a "raw food vegetarian" Swede who spent four months in a "cozy lean-to" in a remote spot on the Big Island.[30]

In Honolulu, journalists and reviewers evinced relative disinterest in discussing Harrison as an African American. Other, more local categories were available to assess her, like "Beachcomber Authoress."[31] Most arresting was her identity as a woman of mixed race. Norwood lauded her as "Part American Negro, Part American Indian and Part just plain American."[32] *Nippu Jiji* identified her as "a woman of negro and Indian blood," while Dew maintained, "Juanita's mixture of blood had created a striking-looking young woman. She wore her long black hair hanging down below her waist"—always adorned with a "bright hibiscus."[33]

(Incidentally, this "young woman" was pushing fifty.) Such references to a "mixture of blood" are entirely absent from syndicated reviews in the continental United States. The difference could be due to the inclination of journalists in Honolulu, Harrison's own emphases there, or both. In Hawaii, announcing blended heritage might have deepened her acceptance in a society in which interethnic alliances were the norm. But perhaps that's making too positive a reading. Garibaldi argues that in the mainstream press, "the particularities of this Black author's lineage were either obscured, or hidden, from white readers," since representations of "interracial liaisons remained taboo."[34] Thus only in Hawaii could such a family story be circulated.

Harrison's connection to the city's Japanese community intrigued at least one journalist, James Hamada, who has a footnote in literary history for writing one of the earliest Japanese American novels, *Don't Give Up the Ship: A Novel of the Hawaiian Islands*. Hamada furnished *Nippu Jiji* with a front-page feature that ran under the headline "Juanita Harrison, *Atlantic Monthly* Writer Now Here, Is Mighty Fond of Japanese Baby." Sedgwick had visited Harrison just five days after the birth of the Tadas' youngest child. Proof of her intermittent efforts to sustain her public exposure and keep her authorship current, the following month Harrison stopped by the offices of *Nippu Jiji* to inform them that at her urging, the couple had changed the newborn's name from Lillian Shizuyo to Lillian Ellery.

Hamada introduces his subject with a masterfully periodic sentence: "When a haole woman loves a cat, or is afraid of a mouse, that isn't news. But when a woman of negro and Indian blood, and a contributor to the highbrow Atlantic Monthly at that, takes a special liking for a newly-born Japanese Infant ... [and] attempts to have the baby named after the editor of Atlantic Monthly, Ellery Sedgwick—That, my friends, IS news!"[35] Whereas discussions of Harrison usually start with something like "Juanita Harrison is a colored woman who ...," Hamada's racial lead is quite different, emphasizing her mixed race and favorably contrasting her behavior to that of white women, whose most trivial emotions and acts, he implies, attract undue attention. Employing the Hawaiian term for white islanders, *haole*, he chose not to present them as racially neutral.

"The feeling was reciprocal," Hamada notes of Harrison's affection for the family. His article, which is subtitled "Here's International Friendship," exhibits a like equity, attending as much to the Tada family as to Harrison. It is illustrated in kind, by both a photo of Harrison posed with two young Tadas in kimono and a close-up of the baby.

FIGURE 5. Harrison with two young Tadas. Source: James T. Hamada, "Juanita Harrison, Atlantic Monthly Writer Now Here, Is Mighty Fond of Japanese Baby," Nippu Jiji, April 22, 1936, p. 1.

One could certainly interpret Harrison's proposal as an impulse to whiten, or at least Americanize, Lillian, supplanting "Shizuyo" with the hoary, patrician "Ellery" favored by upper-class New Englanders. However, this cross-gender name change also indicates the same ambition for the child that her mother voices: "Mrs. Tada expressed the hope that baby Lillian would become a famous editor like Ellery Sedgwick when she grows up. She thinks the child may get a break to become a good writer or a newspaperwoman, if not a famous editor like that of Atlantic Monthly, now that she has been 'blessed' by that famous man."[36] Her parents did not in fact change her name, but as Hamada remarks, it made a good story.

Sedgwick's meeting with Harrison was hastily arranged. Passing a short layover in Honolulu en route to Japan, he and one of his old Harvard friends had been visiting Walter Dillingham when a journalist telephoned to suggest that he call on Harrison. Dubbed the "Baron of Hawaii Industry," as director of Honolulu's largest dredging company Dillingham had overseen the draining of Waikiki's wetlands, and he suppressed Japanese and Filipino labor protests during the 1920 Oahu Sugar Strike. The proceeds from his enterprise funded the construction of the opulent Italianate villa where he received Sedgwick.

Sedgwick's next host and next villa could hardly have been more different. He rolled up to Villa Petit Peep in a ten-car entourage, catching Harrison off guard. In his memoir, *The Happy Profession*, he uses racist language to describe his first sight of her: "'Sakes alive!' I heard in a syrupy gurgle. 'I ain't got a mortal thing on me.' But things were found, the tent flap parted, and out came Juanita, her teeth shining under a carmine bandana, her big eyes bright as blobs of Mississippi molasses."[37] Sedgwick also alleges that Harrison used the N-word to refer to herself. Resurrecting American racial and regional hierarchies, he states that whenever he has troubles with his maid or cook, he thinks longingly of this "Black Pearl among Servants."[38]

We can thus appreciate all the more *Honolulu Star-Bulletin*'s account of the same event, which portrays Sedgwick stepping into Harrison's world and evaluated on her terms. The front-page photograph accompanying the article shows the two of them at Villa Petit Peep, identified as such by lettering on the white picket fence that fronts it.[39] On first inspection, the image may appear to position Sedgwick as authority and judge. Harrison is posed showing him the stamps in her passport, offering proof she has been to the places she claims. The contrast between his natty three-piece white suit and her outfit—a floral print skirt and sleeveless top—makes the encounter look near colonial (the "carmine bandana" of Sedgwick's racist fantasy is nowhere in evidence). Yet Harrison leans over Sedgwick, who sits awkwardly on her fence. The headline, moreover, steers the scene in an unexpected direction in referring to Sedgwick not by name but by Harrison's humorous term for him: "Juanita, Unique Authoress, Meets 'Unseen Sweetheart.'" Sedgwick notes that Honolulu knew him as "the friend of Juanita Harrison," and although the comment is made in the spirit of an inside joke, this was just how he was viewed.[40] Even as he categorized Harrison according to the other Black women he knew—as a servant—Harrison positioned the renowned editor as still another of the enamored, generous, helpful men she regularly met on the road.

Sedgwick's virulent racial views make him an unlikely person to have given Harrison her big break. He almost never published African American writers. Why Harrison? One theory, tendered by Adele Logan Alexander in her introduction to the 1996 reprint of *My Great, Wide, Beautiful World*, is that Harrison's publishers pandered to popular appetite for minstrel performance. Alexander contends that the text "clearly proclaimed its uncelebrated black author as a 'primitive': a lovable yet somewhat clownish Aunt Jemima or latter-day Uncle Remus, whose narrative was readily acceptable to white America as part of a traditional and popular black dialect genre."[41] Some readers did indeed register it that way.

However, as I have argued elsewhere, Sedgwick's work with Harrison extended his long-standing practice of debuting unusual working-class women writers. Starting in 1911 with Mary Antin's *The Promised Land*, throughout his three-decade *Atlantic* career he published the life narratives of women who were located geographically, ethnically, or socioeconomically outside of the mainstream. He dubbed them "Faraway Women," explaining in a *Happy Profession* chapter of the same name, "It was to women then that my thoughts oftenest turned, and a score of lonely, self-dependent histories were woven into the texture of the Atlantic."[42] The group includes a child diarist in an Oregon logging camp; homesteaders in Wyoming, Idaho, and far northern Alberta; and a woman who walked through Alabama for her health. Harrison was the last "Faraway Woman" Sedgwick published before his retirement. She always referred to him warmly; his portrayal of her betrayed her trust.

While Harrison relished the tributes her book drew, she was anxious that its publication might damage her relationship with the Dickinsons. Writing to Foster, she fretted over the possibility that "there would be something in it that wouldn't be just the proper reading" (FL3). The renown it brought mattered

Juanita, Unique Authoress, Meets 'Unseen Sweetheart'

ATLANTIC EDITOR CALLS ON AUTHOR

FIGURE 6. Harrison and Ellery Sedgwick meet in Waikiki. Source: "Juanita, Unique Authoress, Meets 'Unseen Sweetheart,'" *Honolulu Star-Bulletin*, March 25, 1936, p. 4.

less than the social fabric it was fashioned from and potentially tore. Just what did she fear might be improper? Rather than paraphrase her densely expressed sentiments, better to quote them in full:

> Since the Passing away of the gentle and kind Mr. Dickinson I wish nothing from Los Angeles and it pain me when I must write to the family but they are so kind and interested as ever but I am not. I could not open a page of the Book for a month altho. the ones that the Publishers sent me were right at hand it isn't my Book its Mr & Mrs Dickinson its their Kindness and never failing faithfulness to me the 8 years why I wrote each line and it were their own Carefulness that of all those years each letter were saved so my happiness came only through them I care nothing about the rest of the world when it comes to the Book and from the time that the Publishers accepted it I had a heavy heart for fear there would be something in it that wouldn't be just the proper reading and it being Dedicated to the only two people that were ever true and kind to me. Not for them selfs but for their three beautiful Grand children just entering into their theens. So when after Mrs Tufts rec'd the first copy to L.A. and I received this wonderful letter that it was just perfect. for the first time my heart became light. I was glad I sent the joyous answer by the Clipper May 7 reaching them at luncheon May 9 when all in the family enjoyed it and that same evening I ~~wepped~~ wept as I had never done in my life as I am not the one to weap but I felt it and at once wrote to my agent both in New York and San Francisco that I did not want the Book sold in Honolulu. I cared nothing about the money and I also went to the Library and ask them not to take it but they did also I rec'd a beautiful letter from my Publisher saying that in a short time I would not be troubled with publicity but it was that sadness that warned me of not wanting it. I am glad that you and your Children liked My Great Wide Beautiful World. The second Edition are being published now it were a best seller here and I have autographed for many fine People Visiting here. (FL3)

By the end of these musings, she seems to be have written herself past "that sadness," gratified that the book sold so well and was appreciated by so many. She is steadfast in attributing the book to the couple's support, stipulating, "Its their Kindness and never failing faithfulness to me the 8 years why I wrote each line" (FL3). That the Dickinsons motivated "each line" of the book is a remarkably emphatic claim. Of note too is her upending of the trope of the loyal retainer. Here it is the wealthy white former employers whose "faithfulness" is extolled.

These complex feelings were surely behind Harrison's decision to duck out of an event that promised to introduce the "latest member of Honolulu's literary colony, Juanita Harrison, author and vagabond-housemaid."[43] Her appearance at the Little Theater was reported on thus: "A toss of her long locks, a hop, skip and jump across the stage and a hurried greeting to the people of 'My Great Wide Beautiful World' was all that Juanita Harrison, author, erstwhile housemaid and world traveler, would let her public see of her yesterday." The article continues on to describe her exit: "Clad in gay shorts and accompanied by her dog Pluto, Juanita was just about to make her escape through a window when caught by the photographer. With all the reticence of Garbo, the eccentric author refused to divulge either the title or the substance of her forthcoming book. The duration of her stay in the Islands remained likewise a secret. . . . A look of horror came into her eyes when she saw that her secret getaway had been discovered and swiftly, if not quite silently, she fled."[44]

Although the description approaches ridicule as the writer strives for comic effect, it is revealing in respect to Harrison's dislike of publicity as well as intention for a second book. Letters and profiles yield several allusions to the project, presumably a sequel of more real-time travel narrative. Yet that confident reference to "her forthcoming book" notwithstanding, it never came to pass. With Sedgwick in Boston and Morris in New York, Harrison was physically distant from the people who had helped her publish. But more important than her location were the structuring conditions of her authorship and the limitations of the Faraway Women genre. She could publish a book exactly because it was so unlikely that she would, conditions not amenable to sustaining a writing career.

Furthermore, she seemed less than enthused about the prospect, having once commented of her master life plan, "The two Books are not included" (FL2). Like enduring jobs, permanent homes, and long-term relationships, authorship, too, threatened her liberty. Perhaps a mere modest disclaimer, but perhaps a sign of real aversion, another newspaper account of the same event relayed, "She claims that she is not an author, that she is just a maid and that she did not write a book; she only wrote letters to a friend from every place she visited."[45]

Harrison was introduced at the Little Theater event by Mary Ann Tufts, George Dickinson's niece, who was vacationing in Hawaii with his widow, Myra. Tufts was a "book reviewer" at "Home Clubs," as the census put it.[46] She contributed book reviews to southern California newspapers and gave talks at meetings of the Ebell Club, the San Bernardino Woman's Club, and other groups, rather bland offerings along the lines of "The Trend of the Times in Books and Events."[47] Tufts promoted Harrison in Los Angeles as well as in Honolulu.

While Mary Ann Dickinson grew up in Great Falls, Montana, she moved to southern California as a young adult, living quite near Myra and George in Los Angeles. Even after she married, she took long trips with the couple, earning a reputation for "extensive travels in all parts of the world."[48] Thus she knew Harrison when she worked in the Dickinson household. Shortly after Harrison's book came out, *Los Angeles Times* columnist Alma Whitaker reported, "Mary Tufts was bragging on Juanita Harrison, author of 'My Great Wide Beautiful World,' this season's bestseller. You may guess Juanita's glory, when I tell you that Mary was bragging that the lady was formerly colored cook to her aunt, Mrs. George Dickinson—and even that slender link is considered worth boasting about."[49] This white journalist's comments, just a shade short of snide, hint at the social intricacy of Harrison's interracial relationship with the extended Dickinson family.

Less coyly but just as off-key, the *San Bernardino County Sun* recounted that in Honolulu, "Mrs. Tufts has made and renewed acquaintances with a number of writers well known in Los Angeles. Robert Eskridge, Don Blanding, Katherine McFarland, and Ruth McKee ... and other celebrities are summering there. Juanita Harrison, the Negro girl who made such a great success with her book, 'My Great Wide Beautiful World,' and who was formerly a maid in Mrs. Tuft's family, is enjoying the little tent home she bought with her first royalties."[50] The reporting has an odd elision, in that it does not quite identify this "Negro girl ... formerly a maid" as one of the writers "well known in Los Angeles" whom Tufts visited, even though Harrison was far more of a celebrity than small fry like Ruth McKee.

Erroneously, *Honolulu Star Advertiser* identified Tufts as "the woman who first discovered" Harrison. Characterizing Tufts as her "benefactor and correspondent," it alleged that upon recognizing the "real literary value in the quaint letters which the little housemaid wrote her from foreign ports," she had "edited and sent Miss Harrison's travel letters to the Atlantic Monthly."[51] Harrison's editor and agent was of course not Tufts but Morris, and there is no evidence that Harrison even wrote to her. She did however play a role in the book's distribution and publicity, as Harrison informed Foster that "Mrs Tufts rec'd the first copy to L.A." (FL3). The familiarity of the reference—the assumption that Foster knew who Tufts was—confirms their prior acquaintance.

Myra and Mary Ann had embarked for Hawaii on the luxurious *Lurline* in June 1937—sailing first class of course—returning to Los Angeles two months later. Rather surprisingly, while Myra's son, William, had been in Honolulu just the previous month, his stay did not overlap with hers. Nevertheless, indicating the strength of her ties with the family, Harrison spent considerable time with William and his wife. She divulged in a postcard to Clay, her Canadian friend, "The last 10 days have been my happiest. Mr and Mrs William George Dickinson the son of Mrs. Geo W. Dickinson that I wrote all the letters in my book were here. They are so wonderful and 10 years since I saw them my true true friends."[52] Her enduring connections with generations of Dickinsons may have contributed to her settling in southern California when she repatriated in the 1950s.

This must have been a very sweet time for her: ensconced in Waikiki, showered with royalties, and reunited with the cherished Dickinsons. Myra was now on her turf. She and Tufts likely stayed at the Royal Hawaiian Hotel that Harrison knew so well, accorded special services by the maître d', her friend Umberto Tragella. They surely discussed Myra's sorrow over losing George, the man with whom she had gone from small-town youth in Kansas, to a consulate posting in Acapulco, to wealth and influence in Los Angeles—along the way enjoying automobile tours through the East as a young couple, an

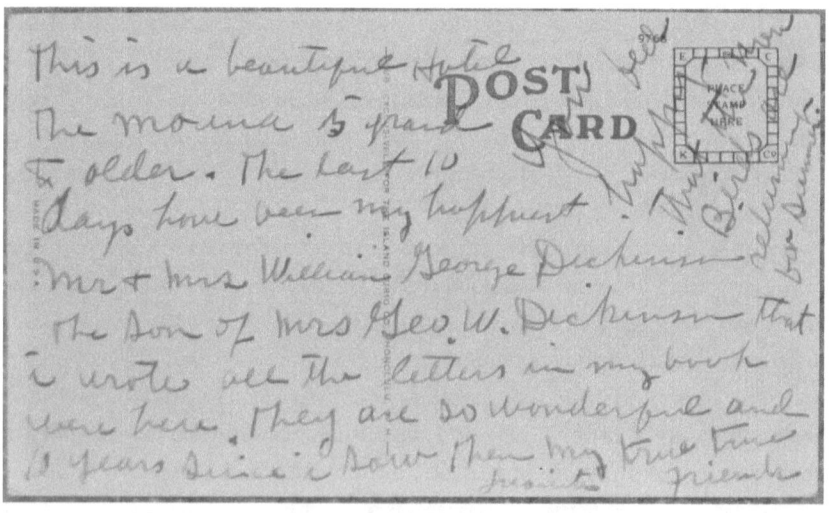

FIGURE 7. Postcard from Honolulu. Source: Juanita Harrison to Mr. Clay, 1937, private collection.

around-the-world trip as a middle-aged one, several early voyages to Hawaii, and other adventures lost to record. But how did the pair comport themselves in public? Harrison, living in a tent on a plot of land she rented for $2.50 a month, had once cleaned the house where Dickinson continued to reside on South Lafayette Park Place, in 1940 valued at $25,000.[53] Oahu was tolerant enough to condone their socializing together, but it is an open question whether Myra was as sanguine, this rich white woman who for decades was assisted by a live-in African American maid, living off income from a real estate firm that pioneered Los Angeles's restrictive covenants.

In that same Lafayette Park house, Myra would go on to host "a series of morning book talks," led by her niece.[54] Doubtless Harrison, a recurring figure in Tufts's repertoire, was on the program for at least one of them. Indicating her local advantage in California, Los Angeles was the only city for which she made the bestseller list in August 1936.[55] Tufts's efforts contributed to her success in the state, as did having a Macmillan agent on the ground in San Francisco (FL3).

The endorsement of Black Angelenos also played a critical role. A different kind of hometown reception played out for Harrison in Los Angeles, one that unlike Honolulu's highlighted her identity as an African American writer. Heralding *My Great, Wide, Beautiful World* as "a book no one can afford to miss," the *California Eagle* instructed its readers to borrow it from Los Angeles's Vernon Branch Library or Helen Hunt Jackson Branch Library.[56] Two months later it announced, "Miss Harrison's book is the first work by a colored author to get on the 'best seller' list of the New York Herald Tribune."[57] The city's largest Spanish language newspaper, *Opinión*, also reviewed the work of this "mujer de raza negra," humorously concluding, "Juanita desconoce completamente lo que es prosodia, lo que es sintaxis y lo que es ortografía. En esto, muchos que escribimos nos parecemos a Juanita."[58] (Juanita has absolutely no knowledge of the nature of prosody, the nature of syntax, and the nature of orthography. In this respect, many of us who write resemble Juanita.) As late as 1959, Ruby Berkley Goodwin included Harrison in a *Los Angeles Sentinel* survey of "California's Negro Authors."[59]

Her most active promoter was likely Miriam Matthews, the director of the Vernon Branch Library who became the leading archivist of the history and cultural production of Black California. Vernon Branch had a large African American membership, and Matthews curated, according to the *California Eagle*, "a section of books by and about the Negro which is a very special attraction to those interested in the subject. Students of the race problem come from all over the city to consult this Negro collection."⁶⁰ Shortly after Harrison's book came out, Matthews organized a summer exhibit about her composed of letters, cards, and photographs lent by Alice Foster. Unusually foregrounding her labor along with her travels, she introduced Harrison as "an American Negro Woman who worked her way around the world in ten years." The photos are not extant, but the preserved captions indicate they were either the same or quite similar to those featured on the book's cover: "The Author at La Riviera, France," "Juanita Harrison in Bethlehem Dress," and "The Author in Turkish Head Dress." Also available for inspection: "Tea Leaves Sent from Burma, India by Juanita Harrison Six Years Ago."

In a brief thank-you note, Matthews assured Foster, "If you could have seen the intense interest that was aroused at the library by Juanita Harrison's exhibit of letters, post cards and photographs you would realize how much we appreciate your great kindness in lending them to us. Although this exhibit has been in the glass case all summer, it continued to boost the circulation of 'My Great, Wide, Beautiful World' until the day it was removed. We still have a waiting list for the book."⁶¹ The "waiting list" suggests the keenness with which ordinary readers responded to Harrison.

Organized by a pioneering African American professional woman, showcasing Harrison's generative friendship with another, and archived by the latter's descendant, the exhibit put Black women's curatorship on display. Matthews stressed Foster's contribution in announcing that the "original letters and postcards by Juanita Harrison" were "exhibited through the courtesy of Mrs. Alice Foster. They were written to her by the author and form part of the book." Her newspaper review of *My Great, Wide, Beautiful World* has a like emphasis. "The letters which make up part of the book were originally written to Mrs. Alice M. Foster," she alleges, and she notes that her family helped Harrison get her first passport.⁶² Even as Matthews gives Foster a shout-out she entirely ignores the Dickinsons, who usually get so much credit.

The discussions of Harrison that circulated in the mainstream press, in contrast, make no allusions to an African American women's support network. None of her publishers and reviewers mention Foster. Although a reader had written to *Honolulu Star Advertiser* to point out its mistake in identifying Tufts as Harrison's editor, the more meaningful, systemic error was the exclusive focus on her white associates.

But of course, Matthews made a mistake too. Harrison explicitly stated that the book's letters were originally addressed to the Dickinsons, not Foster as

Matthews would have it. She also credited Myra with motivating her to write, while Alice seemed to have served more as sounding board than inspiration. Could the misrepresentation have been deliberate on Matthews's part, creating the history she wanted? Clearly she had an agenda of her own.

Living in Honolulu, Harrison was unable to see the Los Angeles display. Moreover, while she did send a friend to take pictures, she did not reply to the letter that Matthews sent her about it. With some pique, the influential cultural leader informed Foster, "My letter to Miss Harrison remains unanswered although I am certain it was received."[63] The reason why Harrison did not write back may well be innocuous. However, she frequently indicated how selective she was about which writers and letters merited a reply. We have also seen how she reveled in her ability to assume manifold guises and, in respect to her employers and suitors, take what she needed without giving back what they thought she owed. Her epistolary neglect thus might have been an act of resistance. Just as she refused to play the role of faithful servant, she may have balked at being made a character in a story purveyed by this champion of Black literary Los Angeles. "Its not easy," she had commented in respect to her work with Morris, "to keep me" (FL1). Harrison continually eludes the narratives that would "keep" her, as construed by employers, editors, readers, and scholars alike.

Nor, of course, would she be kept by any one place. In both popular and critical commentary, Hawaii is identified as the terminus of her extended journey. This is indeed true insofar as her arrival there both fulfilled her goal of circling the world and had her back in US territory. However, Honolulu served Harrison, just like New York, Los Angeles, Nice, Mumbai, Madrid, and many other cities, as not an end point but a resting point. Despite her love of Waikiki, she was always poised to leave. Even before her book was published, she told a reporter that she was soon to head to Alaska in pursuit of northern lights,[64] and in the summer of 1936 she announced, "I like my life Here but will sail for a sea voyage quite soon" (FL3). Being on a remote island chain in the middle of the Pacific Ocean must have heightened her restlessness. "The most geographically isolated land mass on the planet," twenty-five hundred miles from the nearest continent, Hawaii did not afford the casual travel options she had enjoyed the last time she was semisettled, in southern France, where she could venture out to Corsica and take road trips in the Alpes-Maritimes.[65]

She ended up lingering in Honolulu far longer than she had planned. This too was typical. While it can seem as if Harrison was always on the move, had she kept to all her original timelines she would have covered immeasurably more ground. Finally, however, she left Hawaii for Brazil. The voyage followed closely on the final 1939 printing of her book, which may have supplied funds for her passage.

She rode out World War II in South America, having left Honolulu sixteen months before Japan's December 7, 1941, attack on Pearl Harbor. More than twenty-three hundred Americans were killed that day, including the eleven hundred men aboard the USS *Arizona*. Thereafter the territory endured three years of martial law, during which its residents' basic rights were violated. "The Bill of Rights and all civil liberties were suspended in the Islands," Willis David Hoover explains. "People suspected of crimes were tried without representation or hope of appeal." Suggesting the pervasiveness of the new order, even the children were "fingerprinted, duty bound to wear ID badges at all times, never to be caught out of doors without a gas mask."[66] The expansive liberty and ease of movement that Harrison had treasured were shut down. Even the beach was cordoned off.

She must have found it harrowing to read about the attack from afar in the Argentine resort town where she was living at the time. She may well have known some of the victims, mourning them even as she worried about the fate of her Japanese American friends. While for all the islands' residents, life was transformed, the Japanese bore the brunt of it. Approximately 2,270 Japanese residents out of a population of 137,000 were interned, most at the Honouliuli Internment Camp. This number was a cruel one, but proportionally much lower than in the western United States. Since the Japanese comprised the largest ethnic group in Hawaii, local authorities understood that its economy could not tolerate the imprisonment of so many workers. Jane L. Scheiber and Harry N. Scheiber explain that rather than mass internment, the detainment strategy in Hawaii was to target "the elite of the Nikkei community, explicitly designed as a way of undermining the various sectors of community life."[67] The Tadas were not of that social stratum, and there is no evidence that they were held. Yet like all their countrymen they faced many challenges.

During the war the family was no longer at the Ohua Avenue property where Harrison had lived, but rather at an apartment on 2564 Lemon Road. The move intimates some hardship or at least upheaval. Not long afterward they moved into a large house on Duval Street, also in Waikiki, where the family continued to live up until Kuni's death in 1975. Indicating that they were both weathering the ordeal and displaying their American loyalty, a 1942 issue of *Nippu Jiji* announced that the newspaper had helped four Japanese residents, among them the Tadas' youngest daughter, purchase a United States war savings bond.[68]

The later military service of their son George Yoshimato Tada testifies to the family's patriotism. Like many young Japanese American men, George enlisted after graduating from high school, and he went on to have a distinguished career as a decorated officer.[69]

Lillian Shizuyo Tada, the baby Harrison proposed renaming after Sedgwick and the beneficiary of the twenty-five-dollar bond, is the protagonist of a quieter American success story, as dictated by the episodic record of her youthful activities. At Kaimuki High School at the edge of Waikiki, Lillian served as a "song and cheer leader," as a "sergeant-of arms" for the YWCA youth chapter, and, one Christmas, as a member of the "Santa Claus Welcome Committee."[70] In March 1953, age sixteen, she sang in the chorus of the exuberantly composite "Hapa Haole Follies." The "benefit show," readers of the *Honolulu Advertiser* were promised, offered

> Six acts including a Harlem "Boogie-Woogie" with Negro music and dances; "Hillbilly" with cowboys and dogpatch characters; a Puerto Rican act featuring drummer Thomas Valentine in Calypso-Trinidad music; and a dance and vocal by Mathilda Kipihea and George Cabang, respectively. A "Joe College" act will feature chorus girls and George Cabang. There will also be a version of "Hawaii Calls" with Frederick O'Riely as Webley Edwards. The arrival of the Lurline will be enacted by Samoan dancers and others. Another act, "Glimpses of the Far East," will star Marjorie Lynch in Japanese, Chinese and Siamese dances.[71]

After graduation Lillian was employed at Sears Roebuck and Company, and she was nominated for the estimable Sears Citizen of the Year award.[72] While Harrison's life journey was uniquely scaled, it is worth remembering that she was surrounded by friends and associates who more privately and conventionally— but just as earnestly and diligently—themselves navigated shifting social terrain.

Even with all the disruption of martial law, Honolulu still reported on Juanita: A November 1942 *Honolulu Advertiser* notice revealed that this once "familiar figure on the beach" was now in Mar del Plata, Argentina, adding the romantic detail that ocean waves delivered the firewood with which she cooked on the rocky shore.[73] Perhaps memories of her carefree days in Waikiki, and the assurance that she continued to live in much the same way in another seaside idyll, offered the beleaguered citizens of Honolulu a vicarious sense of freedom and delight. Their ongoing interest in her doings could have smoothed the way for her late-life return to the islands. The next chapter discusses Harrison's experiences in South America, but we will revisit her in Hawaii at the end of this book.

CHAPTER 9

Pan-Hispanic World

Juanita Harrison had no Hispanic heritage. The popularity of the name Juanita in the United States in the late nineteenth century—as well as others such as Isabella, Gloria, and Ramona (the vogue for the latter triggered by Helen Hunt Jackson's eponymous blockbuster)—points to the indelible global imprint of Spain. The Spanish empire was one of the largest in history. Although the loss of the American colonies had long since precipitated its decline, its once vast reach and enduring legacies made for countless places where Harrison was served by her cumulative knowledge of Hispanic culture. Throughout her life, she navigated a pan-Hispanic world, moving from Spanish-legacy regions in the United States to Cuba, Spain, and South America, and then back to the United States. And being called "Juanita" gave her a boost. Of the children she cared for in Barcelona, she remarked, "They like me because my name are Familer to them" (174), and she described the happy astonishment of the vendor in Seville with whom she shared a riverside breakfast: "He ask me my name and when I said Juanita he was so surprised he leaned back and the jug rolled from under him I ask him if he didnt like it he said of course it was the one name he did like" (207). Through travel and expatriation she recuperated the Spanish identity of "Juanita"—no longer just a trendy American name, but an actual Spanish one, too.

This chapter departs from a strict chronological progression to recursively investigate Harrison's experience around the pan-Hispanic world, building context for understanding her South America residence, that major but sparely documented life stage in her fifties and early sixties. Her first Latin sojourn, of a kind, was in Florida. Afterward she lived in Havana and US borderland locations. Her world voyage took her to different regions in Spain, and after some hard traveling she recovered in its former outpost in Asia, the Philippines. 1940 marks the start of her immersive thirteen-year interlude in South America. She spent six months each in Brazil and Uruguay, traveled in Peru, Chile, and Bolivia, and lived for nearly a decade in Argentina, first in a seaside resort town and then in the capital. While geographically far-flung, these places are bound (save Brazil) by shared language, histories, and customs, comprising a persistent Hispanic presence that threaded through much of her life.

Harrison's application to renew her passport in Buenos Aires lists Florida as one of her former states of residence, with its position in the series suggesting she went there after New York. Perhaps she alighted at St. Augustine, given that she remarked of Geneva, Switzerland, that "it make you think of St. Augustine Fla. they have the buggies with the Fring Tops and Two rivers run through the City" (40). Venerable St. Augustine, with its coastal location and storied past, was exactly what she liked. Founded in 1565, aside from a brief British interlude, the settlement remained under Spanish rule for over two centuries, until the United States took possession of Florida in 1821. Some hundred years later, capitalizing on its Spanish heritage and fine climate coupled with a modern railway and new luxury hotels, St. Augustine was marketed as a winter resort.

Florida was the point of departure for Harrison's move to Cuba, reached in scant hours by ferry. By the late 1910s, the Peninsular & Occidental Steamship Company ran two weekly trips from Tampa to Havana, ports soon joined by Miami and Key West.[1] The ferries were coordinated with rail itineraries, the linchpin of a concerted campaign to attract American visitors. Within forty-two hours of leaving New York's Penn Station, passengers on the *Havana Special* could cross 1,596 miles by rail and ship to be sequestered in their hotel rooms in Havana.[2]

Cuba was the site of many firsts for Harrison: her first international destination, first tropical residence, and first Spanish-speaking country. During her time there, she learned Spanish well enough to be told years later that she "spoke spanish just like a Cuban" (178)—a dialect known for the dropped consonants that make for "relaxed pronunciations," as one Spanish language blog puts it.[3] Inexplicably enough, her time in Cuba is contiguous to a period of residence in Canada. Between 1917 and 1922, she lived in both countries, and the 1937 letter to her friend Clay in Vancouver, British Columbia, indicates a likely residence in that northwestern city, close to three thousand miles from Havana.

Her keeping in touch with Clay all that time shows she made some friends in Canada. However, all we learn from the book about her experiences there is that she worked for an Englishwoman and saw Niagara Falls. While in the nineteenth century, Canada was a site of inspiration and real harbor for fugitive slaves, Harrison presents it quite neutrally, as just another place of work and tourism. Yet in contrast to Canada's meager showing in the text, she mentions Cuba a full nineteen times. The island nation had a deep impact on her, both during her actual stay and through its effect on later life choices.

Cuba was a triangulation point of European, American, and African conquest and trade. Revolutionaries had succeeded in winning independence from Spain in 1898, aided by the United States. While Cuba's Republican Era began in 1902, the United States retained partial military and economic control until 1934, and Harrison's residency coincided with a heady capitalist era. The advent of World War I sparked a tourism boom, as Americans cut off from Europe sought out alternative vacation destinations. Those with vested interests in Cuba's nascent tourism industry seized the opportunity. Louis A. Pérez states in his history of modern Cuba that "as early as 1914, U.S. railroad interests, shipping executives, hotel operators, and retail merchants formed the Cuban Commercial Association for the purpose of 'attracting foreign travel.'"[4] By 1916 there were forty-four thousand annual visitors to Cuba, and within twelve years that number had doubled.[5] A 1920s postcard promoted it as "So near and yet so foreign."[6] It was "precisely the juxtaposition of the foreign with the familiar," Pérez clarifies, "that was at the heart of the Cuban appeal to North American sensibilities."

With its accelerating economy, Cuba was not only an alluring place for Harrison to visit but also a practical one to find work. Similar to her later experience on the French Riviera, that she was there just as its population of expatriates and holidaymakers surged made for a surfeit of service jobs to choose from. A host of economic migrants gave her company. American bartenders, for example, flocked there during Prohibition, to continue practicing their profession in a country where alcohol remained legal.

Havana was dubbed "Paris of the West Indies," "Paris of America," "Paris of the Antilles," "the Nice of the Atlantic," "the Cuban Riviera," "the Riviera of the Caribbean."[7] What a difference from rural Mississippi, Harrison's home less than ten years earlier! She may have been a witness to Cuba's "Dance of the Millions," its dizzying, short-lived sugar bubble. Just as American tourists could no longer travel to Europe, European beets could no longer reach America. Cuba thus became a major player in the industry, sparking instant fortunes. Kevin Grogan describes the "dreamlike atmosphere" there during the first months of 1920.[8] Yet sugar prices dropped just as abruptly, triggering the failure of many of Cuba's biggest banks. Hopefully Harrison's savings were not on deposit in one of them, a calamity she would experience soon enough in Denver.

While Havana had a reputation for "vice," for drinking, gambling, and sexual license, in her usual way, Harrison undermines those associations in her text.[9] Her allusions to Cuba conjure an eminently wholesome lifestyle. She worked—of course—for an Englishwoman and at least one Spanish employer. She traveled—of course—savoring Cuba's country hotels, later a point of reference in the Middle East. She adopted the habit of never leaving a church without praying, and her regular mass attendance laid the foundation for her eventual conversion to Catholicism. In her book she positions herself as a non-Catholic, observing, "I must have a will of my own otherwise I would have become a Catolic. I am always meeting ones that try hard to get me in" (234). By the mid-1940s, however, she did indeed identify as Catholic.

Cuba had a very large African population, descendants of the thousands who had been forcibly brought from Africa to work its sugar plantations. Emancipated slaves fought for the nation's independence from Spain, a full 40 percent of its military force, and the next generation continued to toil in the cane fields. However, Harrison never mentions Afro-Cubans. Instead, Cuba helped her look ahead to Europe. The "Riviera of the Caribbean" oriented her to the French Riviera, as we see in her observation in Boulogne-Sur-Mer that "this are like a Cuban town" and "the sea front are very much like Havana" (15). As more material preparation, while living on that Spanish-speaking island, for a full year she took classes in *French*, indulging her penchant for imagining herself into a new place even as she was immersed in her present one.

Havana also paved the way for Honolulu, which she first visited several years later. Cuba and Hawaii had very different populations and colonial legacies, and thanks to its missionary past, the latter's image was far more sanitized. Nobody described Honolulu as "one of the world's fabled fleshpots," as they did Havana.[10] Yet other likenesses are evident. Both Hawaii and Cuba were occupied by the United States, both had plantation economies and pernicious ethnic hierarchies, and both had a mushrooming tourism industry that promoted enchanting yet anodyne experience. Her employers' garden in Waikiki, Harrison remarks in her book, had "a pipia fruit tree banna tree and another fruit tree that have a fruit that are much liked in Cuba as a drink The Mr. and Mrs. didnt know what it was and let it rot until I came. now they cannot get enough" (313). Along with proving that a maid could be more worldly than the people she served, the minor incident suggests how Cuba paved the way for other tropical sojourns.

That she had once lived in Cuba helped Harrison control her public image. As discussed previously, she signed the guest register at Kīlauea Volcano House as "Mrs. Juanita Harrison, Habana, Cuba," and in Nice, she outright claimed to be Cuban, explaining she did so because "the American and English drink and gamble so much" (24). Latin America offered her, along with some of her African American contemporaries, ways to dissemble about her national origins and perhaps about her race, too.

Framed by briefer stays in Florida and Louisiana, her residence in Cuba was succeeded by those influential years in Los Angeles. The American sojourns sustained her Hispanic exposure. Los Angeles had an enormous Mexican population, while New Orleans can be conceived as "part of a greater Spanish borderland," to quote Kirsten Silva Gruesz, mediating "between the Anglo and Latin worlds."[11] We have to guess, though, about the nature of her Latin contacts in these places, as Harrison offered no clues. Yet not too long afterward, she began her most closely documented Hispanic period, her year and seven months living in different regions of Spain.

While France was Harrison's original European dream destination, she grew to love Spain, too. After France it was the European country where she spent the most time. Starting in October 1929 just as she lost her investments in the stock market crash, she moved between Valencia, Barcelona, Madrid, Seville, and the village of Molina del Rey. She assessed Barcelona as lively and picturesque, Madrid as modern but studded with pockets of tradition, and Seville as "a great place for learning everything thats Spanish" (205). Molina del Rey was arrestingly rural: unpaved streets, delivery by burro, and the morning chore of sweeping up horse dung.

Shortly after first arriving in Spain, Harrison reported that although she "had forgot much of my Spanish," she was "picking it up very fast" (183). Her Spanish proficiency came to vie with her French, so much so, she revealed, that when she met Frenchmen on the streets of San Sebastian who "want to speak French I tell them I know Spanish very well" (235). Her language ability made it far easier to get hired, by both native Spanish speakers and American employers who relied on her for translation. Although she contemplated alternatives such as selling newspapers or working in a sardine-processing plant, all her jobs were in domestic service.

My Great, Wide, Beautiful World shows Spain to be a nation in transition. Harrison's very presence there proved that it was moving into a modern era, attracting itinerant women workers; she refers in passing to the Girls Friendly Society in Madrid, still another charitable organization that offered migrants practical aid coupled with Christian uplift. Nonetheless, many older customs still prevailed, and she consistently portrays Spain as antiquated.

She found its traditions delightful when they took the form of spectacle. She comments, for example, that she liked Spanish parades more than Cuban ones, because horses, not automobiles, predominated. She was frustrated, however, by what she saw as hide-bound customs, especially as practiced in private homes. "They have many ways of the East" (192), she observes, and elsewhere elaborates, "They have many old ways of during things the Spaniards are like the Indians their coustoms are to old The laundress get 12 cents an hour and wash sheets in cold water" (176). The backwardness she attributes to Spain manifested itself in ubiquitous poor labor conditions. Spanish employers resisted modern household technology to the extent of vetoing mops. But if she was taken aback by such mandates, she was appalled by the pay. Of the maids in Seville, she marvels, "how the Girl can sing on top of those small wages and hard work" (204). She was also startled by Spain's low literacy rate, remarking "it seem so strange a country so near the other European Countrys ... the most common question you hear is can you read" (202). She herself could answer that question with a proud affirmative—in several languages.

Her rebellion puts her American identity on display. In respect to her housekeeping job at a children's home in Barcelona, Harrison reports, "I have caused some changes the Family eat early and quicker" (no small feat in Spain), and she had them replace cookware she deemed unsanitary. "I can teach a Spanish maid more in a minute than she can teach me in a week" (175), she boasts. She refers here to housekeeping methods, but she was also intent on "teaching" a change in perception, as when she attempted to convince a young working mother, "a pretty blonde girl of 20," of her rights. She recalls,

> She so young and pretty I felt so sorry for her. I went out and she was surprised to see me with a hat and gloves. She said I truly must be rich if not I would not ware a hat I told her in my country, France and England everybody wore hats but she couldnt see it. she said she like so much to ware a hat but she being a servent could not I told her if she wanted to ware a hat to ware it being a servent made no differents but she could not see that. (185–86)

Insisting that being in service should not shut down one's autonomy, she uncharacteristically identifies the United States as "my country."

In that same city, the 1929 Barcelona International Exposition was the official face of modern Spain. Twenty European countries were represented at its sprawling hillside site. It ran for nearly nine months concurrently with the Ibero-American Exposition in Seville, devoted to Latin America, which Harrison also visited. As usual she had planned her stay to coincide with a grand national event. "I made a good move when I choosed Spain at this time of the Exposition" (174), she states. Typical of world fairs, the Barcelona Exposition celebrated colonial legacies. For Harrison personally, the lead flavor was entertainment. She recounts, "I went to the Fair again and meet several Bombay Ceylon and the English Gentleman that mananges the Oriental Theatre of 7 Ceylon dancers and a good Snake Charmer I went in Free. I laughed so when the Ceylon dancer that look so much like a Girl told me how a Spanard tryed hard to make love to him" (173). Having testified to the fair's abundance, in the next sentence she suggests that the city of Barcelona itself offered comparable variety: "I can get everything in two blocks of my street from a good plate of Itialian spaggetti to a Turkish bath" (173).

Befitting the nation's capital, her most up-to-date Spanish sojourn was in Madrid. Her American employers, Mr. and Mrs. "B.," had placed a Help Wanted notice for an English-speaking nurse in Spain's leading daily newspaper, *ABC*. Harrison introduces the position with the couple and their two-year-old daughter, Joan, by affirming, "I owe thanks to Madrid because the secon day there I got a good place and good pay and keep it until the day I wanted to leave" (221). "Here are many Americans Familys their Husbands are here with the Telephone company" (212), she reports, and presumably Mr. B. was among them. Just as she personally brought modern housekeeping methods to private households, American corporations brought modern technology to Spain.

Indicating their high status, the B.'s resided in a suite at the Petit Palace Savoy Alfonso XII Hotel, one of Madrid's top luxury hotels. Their quarters overlooked the gracious Parque del Retiro, which reminded Harrison of Central Park in New York. Indicating her niche service role as a Spanish-speaking American, she explains, "They do not speak Spanish so are delighted to have me" (211). She also did some translation of English into French when Mrs. B. shopped at pricey boutiques. Paying her well above the going rate, the couple also covered her rent so that she could live apart from them on an "ancient old street with all Spanish" (217) near the Plaza Mayor. From the window of her room she could see the dome of the Royal Basilica of San Francisco el Grande.

While Harrison was fond of little Joan, she found the young Mrs. B. a trial, her provincialism at odds with her own worldliness. She complains, "I had to do the thing I so much dislike I had to go to a French shop to talk for the Lady I am with. They are to cheap to be true Americans" (216). In her view, "to be true Americans" meant spending freely. She observes, "This is the first time She have been out of America and expect to find the same make here as there yet they have nice and good things in Spain" (218).

They also clashed over childrearing. Joan was subjected to long shopping expeditions, and Harrison notes that the Madrileños pitied the overdressed child: "They look at Her so hard and tell Her she will sure keep warm. They say she is a Poor Little Darling and ask me if anything is the matter with her feet" (212). The testy relationship reached a seeming impasse when Mrs. B. insisted she "give the Baby anemener" (213). Upon this she quit, but Mrs. B. lured her back with higher wages and shorter hours. If she could last until June, Harrison reckoned, she could afford a trip to northern Europe.

The two women did, however, enjoy some less transactional good times. Showing how easily power dynamics could shift, Harrison taught her boss flamenco and basic Spanish. She also translated the "many loving words" men said to her on the street: "She doesnt understand a word, and when I tell Her She is so surprise. She has Beautiful blue eyes" (212). The fact that those beautiful eyes were blue seems to be posited as further evidence of her naiveté, the guilelessness of the privileged white woman.

The most intriguing Madrid episode is perhaps the outing Harrison made with Joan to the Prado Museum, scant blocks from the family's hotel. She recounts, "Joan and I Visited the first floor of the Museo del Prado One a Painting of King Philip IV and His Queen and Two children are the only one in the room the window are half open so that the light fall on the canvas the only light in the room on the wall are a large mirrow one can see it in it. and make it very real" (217). Diego Velázquez did most of the portraits of King Philip IV and his family, and she almost certainly describes his famous court portrait, *Las Meninas* (The ladies-in-waiting).[12] The massive work portrays Philip's oldest child, five-year-old Margarita Maria, flanked by two attendants amid a larger entourage. One of the women gazes at their charge, but the other meets the viewer's eye. The king and queen are visible in a mirror hanging on the wall, keeping watch over the painter as he fulfills his commission. Velázquez painted himself into the scene, and like the princess and her attendant, he too looks out toward the viewer. Michel Foucault, as a way of introducing the themes of *The Order of Things: An Archeology of the Human Sciences*, surveys these exchanges in the painting: "A mere confrontation, eyes catching one another's glance, direct looks superimposing themselves upon one another as they cross.

And yet this slender line of reciprocal visibility embraces a whole complex network of uncertainties, exchanges, and feints."[13]

We can wonder if Harrison, gazing at Las Meninas, reflected on the commentary it makes about the relationship between childcare, service, patronage, surveillance, and art, the "whole complex network." The thematic connections to her own experience are manifest. Author of a book composed of letters written to employers turned patrons—letters that routinely question the judgment and authority of the people for whom she worked—she resembled both the maids and Velázquez himself. In the company of her young charge, she contemplates a tableau of a rich, adorned, privileged yet constrained girl and her relationship with her caretakers, with whom the child is in far closer contact than the powerful parents who oversee them. The uncharacteristically specific commentary about the lighting—"the only light in the room on the wall are a large mirrow one can see it in it. and make it very real"—reads as a casting about for a way to articulate the effect of doubled perspectives, which enhanced the sensation of it being "very real."

After seven months of work and travel, finally Harrison left Spain, well ahead of the cataclysm of the Spanish Civil War. Especially captivated by Seville, she had to force herself to go. Her residence in that nation, as we know, was succeeded by a much longer one in France. Why did France win out over Spain? Partly the choice was contingent: During her earlier stay in France she had made good contacts, including the Morrises in Paris and the Roses on the Cap d'Antibes. Crucially, wages were higher, especially in American expatriate households. The bohemian society that developed on the Riviera appealed to her, too, much fresher than what she could find on the Iberian Peninsula. When she next lived in a Spanish-speaking country, it would be on a different continent. In the intervening decade she published her book, whose cover, studded with drawings of what look to be flamenco dancers along with a photograph of herself gripping a guitar, captures her love of Spanish culture.

Obviously Harrison's years in Cuba and Spain helped her imagine South America as a potential home. However, her late-life migration may have had another, less evident, impetus—the weeks she spent in Kobe, Japan. A major port, Kobe was a gathering point for government-sponsored emigrants headed to South America. Brazil was by far the leading destination, and in 1933 and 1934—the exact time Harrison was there—emigration from Japan to Brazil was at its peak.

In Kobe, she lodged at the YWCA on Yamamoto-dori. The avenue, a major east–west thoroughfare, runs along the base of the Rokko mountain range.

She was just a stroll away from the rhododendron-lined trails that thread the hills, the teahouses with their views over the city and the sea beyond, and the 130-foot Nunobiki waterfall celebrated in Heian poetry. As I know from having lived there for ten years myself, the key features of the neighborhood are little changed from how Harrison described them.

Less picturesque, a walk due east along busy Yamamoto-dori takes one to Kobe's former emigrant center. Although now a museum, in the 1930s the facility lodged, examined, and instructed the outward-bound men and women from across southwestern Japan who gathered in Kobe prior to embarkation. It was there that their documentation was processed and their health scrutinized. They also received cultural briefings and language lessons, in not only English and Portuguese but also Japanese, to further the government's goal of founding a new Japan in the Americas. Putting Harrison in even closer proximity to the migration enterprise, Kobe YWCA assisted in these efforts. Local middle-class women could take classes—meeting up for tuition in sewing, cooking, ikebana, and English in the cottage annex where Harrison slept—and it received the occasional foreign visitor like herself. Yet the Kobe Y was also a cog in "the emigration machinery," to use Jefferey M. Sellers's metaphor, in running English-language courses expressly designed to prepare women emigrants for the challenges to come.[14]

Hawaii and California were once the leading destinations for Asian workers, emigrants like Harrison's hosts in Honolulu, the Tadas. However, as of 1924 restrictive US quotas diverted the flow southward. Between 1901 and 1940, 40 percent of the Japanese who settled outside East Asia and the Pacific went to Latin America.[15] In contrast to European emigration, which was driven by individual initiatives, most Japanese emigration was state funded. This policy made for unusual settler dynamics, as did the contrast between Japan's economy and that of its target countries, with migration "from a relatively rich, industrializing country in the North to poorer countries in the South" reversing the standard trajectory.[16]

On their departure day from Kobe, the wayfarers walked hundreds strong from the emigration center on Yamamoto-dori, descending to the city streets downtown and then onward to Kobe Port. The majority would never return to Japan. Perhaps Harrison saw the stirring sight, or even spoke with some of the people preparing for this profound life change. Did it sow seeds for her future move? Little did she know that only six years later she would follow them, although, typically, not through a federal program but on her own initiative.

Long before she reached the shores of Brazil, Harrison had begun to conceive of herself as a citizen of the Americas. Living in Cuba and Spain schooled her in the southern continent's dominant language and offered previews of the cultural environments she would enter. Meeting its citizens in Europe did the same, along with fostering an expanded sense of American identity. Her very first job in England was "with a South American Lady," which she got because she "had worked in Cuba and learned Spanish" (5). Later she lived in the countryside with a remarkable Spanish-Peruvian family, whose mobile history attests to the ongoing circulation between Spain and its former colonies, migration and back migration. Harrison explains, "La Sra. were born in this house Built 85 years ago. . . . Her husband were Born in the same Town and went to Lima Perue when a young man and became very rich and he came back after he lost his first wife and married this Lady and she gave him a child every year for eight years four boys which died and four girls all the children were born in S. America they came to Molins del Rey 6 years ago to put the girls in School" (179). She also discloses, "The Girls love their home in S. America and often say to me we all are Americans" (181). She says nothing further about this affirmation of shared American identity, but it made enough of an impact for her to record it. We can imagine that due to her own serial migrations, she could well recognize the truth of their claim.

An equally intriguing exchange took place en route to Burgos with the family she met at the massive Renaissance royal complex, El Escorial. She relays, "They ask me am I argentina" (233). The query was likely prompted by her unusual accent, even though the intonation she picked up in Cuba would have little resembled Argentina's Italianate dialect. Her reply: "But I said no American del Norte" (233). Even as she denied being Argentinean, she posited an equivalence between South America and the United States, rendering the latter "America del Norte." Her recording a portion of her answer in Spanish confirms her affiliation, and she goes on to praise the "good spanish bread."

Harrison sailed for Brazil in 1940 at the close of her four-year residence in Honolulu. She had renewed her passport on December 28, 1939. She probably did so in San Francisco, given that one month later she got a visa for Brazil at the Brazilian consulate in that city. Auguring well for the venture, her visa photo shows her looking vital and robust, a lei draped around her neck.[17] She was fifty-three. Surely it would have been far easier for her just to stay on in Honolulu, but she was keen to explore a brand-new continent—her zeal for change and major enterprise unabated, her thirst for discovery unslaked. Her carrier was likely the Pacific Argentine Brazil, a major line with sailings from San Francisco to a number of Brazilian ports, chief among them Rio de Janeiro.

She must have felt a sense of homecoming on reaching South America, its bold American character infused with eclectic European legacies.

While geographically removed from World War II, South America was profoundly affected by it. Nations were compelled to shift political alliances, and they became more economically dependent on the United States. The impact was also social, as some countries received thousands of European refugees along with a considerable number of Nazi war criminals, especially Argentina. At the time its tourism industry was still in its infancy. Collectively, Latin America and the Caribbean received just 1.3 million foreign visitors in 1950. By way of comparison, in 1980 that figure was 18 million.[18] As usual Harrison was in the vanguard.

Since the mid-nineteenth century Brazil had had a reputation among middle-class African Americans as being especially welcoming, an expectation that Harrison may have shared. Valerie Popp charts their perception of the nation as cosmopolitan and inclusive, largely due to its reputation as "a locale where class is more significant than race."[19] She summarizes the views of the race theorist Martin Delany, that in Brazil, "a black immigrant could fashion a free self without having to abandon an American identity."[20] During her time there, Harrison would have

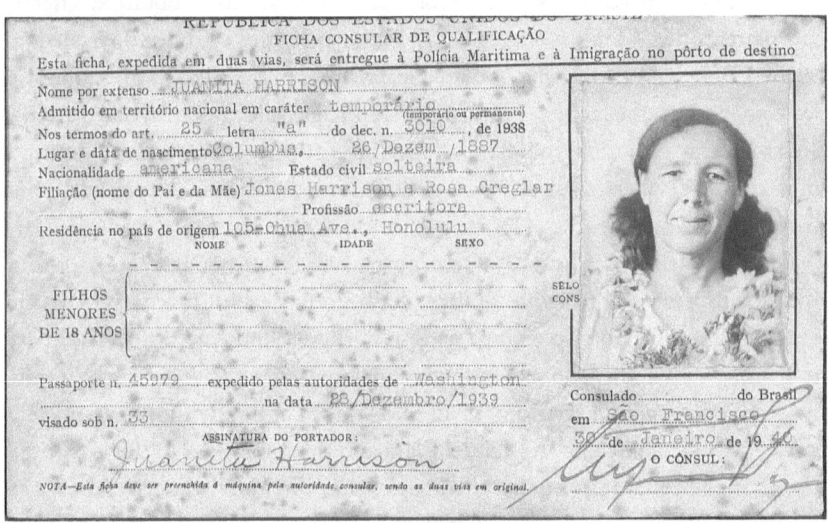

FIGURE 8. Harrison's visa for Brazil—the photo evidently taken while she was in Hawaii. Source: Juanita Harrison, Brazil visa, December 28, 1939, Brazilian Consulate, San Francisco.

learned the difference between Lusitano and Hispanic culture, a distinction to which—judging by her comment in Honolulu, "when I feel Spanish There are the Potegeise" (314)—she had previously been oblivious. Brazilians would have let her know posthaste that "the Potegeise" were not "Spanish."

The 1950 passport application at the US consulate in Buenos Aires, which tells us so much about her movements across North America, is just as valuable for recovering her movements across South America. It attests that she remained in Brazil for half a year, followed by another half year in Uruguay and a year and eight months in Argentina. The first Argentina interval, we learn elsewhere, was passed in the small town of Mar del Plata. Three months in Chile bridged the time between Mar del Plata and the long residence in Buenos Aires that began on January 2, 1943. Her passport had expired in January 1948, after which she likely remained in Argentina, up until she renewed it to go to Bolivia.

Although she must have held various jobs in these countries, she surely traveled widely, too. Brazil, Argentina, Uruguay, and Chile were all popular destinations for American tourists, and if past is prologue, Harrison's travel itinerary was quite standard. She may have enjoyed the Copacabana and Ipanema Beaches in Rio de Janeiro, marveled at Iguazú Falls from viewing points on both sides of the Argentina-Brazil border, and toured the colonial neighborhoods of Montevideo. Excursions to the Andes were a likely highlight in Chile. She stayed in inexpensive guesthouses and ate local food—steak, corn, potatoes, cheese, chocolate, coffee—and in Argentina at least, quantities of dulce de leche and gelato. Outside of Brazil, she operated exclusively in Spanish and became still more adept with the language. As a light-skinned Black woman, while traveling she faced less prejudice than she would have in the mid-twentieth-century United States, where segregationist policies forced African Americans to plot to get rooms and rides.

Directed by her own unrecoverable logic, having launched her extended tour of the continent from Brazil, a country of astounding ethnic diversity, Harrison went on to settle in its two whitest nations, Uruguay and Argentina. The query in Spain, over whether she was Argentinean, forecasts a future in which she did, in a way, become an Argentine, as all told she spent nearly nine years in that country: a year and a half in Mar del Plata and the remainder in Buenos Aires. The pairing is characteristic, as throughout her life Harrison cycled between world capitals and seaside tourist towns.

We know she was in Mar del Plata in 1942 because she wrote to Macmillan in New York to inform them of her whereabouts. Her habits of international correspondence and professional networking had continued. They saw the update as newsworthy enough to pass on to the *New York Times*, which informed its readers, "A recent postcard from Juanita Harrison to her publishers, the Macmillan Company, discloses that she has been living in Mar del Plata, Argentina. She says she does her cooking on the rocks, with the waves bringing in the firewood. Miss Harrison is the author of 'My Great, Wide, Beautiful World.'"[21] A version of this notice followed in the *Honolulu Advertiser*. A good six years after the publication of her book, New York and Honolulu were still interested in Harrison's whereabouts, and they were still intrigued by her lifestyle.[22]

Most of the permanent residents of Mar del Plata were recent immigrants from Europe, especially Sicily, Italy, France, and Spain. Fronting a chain of beaches on the Atlantic coast, Mar del Plata had been a thinly populated ranching, farming, and fishing community before the Buenos Aires Great Southern railway line reached it in the late nineteenth century. The new rail service triggered an influx of settlers, and in the 1920s upper-class Argentineans began to build grand summer homes, enjoying the beaches from November to April. The completion of National Route 2 in 1932, replacing the dirt road that had required almost two days of travel from Buenos Aires, ushered in large-scale tourism. Making for an infusion of visitors and jobs, the seafront Casino Central, modeled after the Hôtel du Palais in Biarritz, opened just two years before Harrison moved there. She might have had a position at the Hôtel du Palais or another fine hotel. More likely, she worked at a private household in an upscale neighborhood such as Los Troncos, known for its leafy boulevards and gracious estates.

Yet eventually she succumbed to the lure of the capital. The seven years Harrison clocked in Buenos Aires make for one of her longest residencies anywhere. As her world voyage drew to a close, she had commented, "The way I feel now I would hate to just stop in one place for a whole year or more" (307). Such insistence on perpetual change is characteristic. Note, however, that she gives herself an out with the qualifier "the way I feel now," allowing for the possibility that her feelings might change. Indeed, later in life she did start to settle down, with greater stretches of time between her moves.

Buenos Aires's nickname, "Paris of Latin America," most directly refers to its neoclassical architecture and urban plan, modeled after the Hausmannian renovation of Paris: the city's symmetrical grid of broad avenues and attendant blocks of stately buildings, parks, squares, and monuments. Yet the moniker has a racial dimension, in that it signals the ethnic identity the nation claimed for itself. Its leaders aspired to form a "European" nation—that is, a white one.

Officially, Argentina is 97 percent white. A hoary popular saying, repeated by its former president Alberto Fernández as recently as 2021, contends that "Mexicans descend from the Aztecs, Peruvians from the Incas—but Argentinians descend from the ships."[23] According to this formula, the population is not Indian but European, and Africans do not even figure in it. National spokesmen facilely claimed that military conflicts and disease caused the wholesale disappearance of the descendants of enslaved Africans. In fact these Black Argentinians were absorbed into the white population. The government offered no other option: In 1890, the category of "African" was removed from the national census, restored only in 2010.[24]

In her 1946 memoir *American Daughter*, Era Bell Thompson recalls, "Ever since reading Juanita Harrison's Great Big Beautiful World, I had been threatening to go around the world, too, had literature on European bicycle tours, rates for tramp steamers, catalogues from the universities of Hawaii and Alaska."[25] These dreams prefigured her extensive travels as long-time editor and foreign correspondent for *Ebony* magazine. In that capacity she went to Argentina to research a feature article, "Argentina: Land of the Vanishing Blacks," with the stated goal of discovering "what happened to Argentina's involuntary immigrants, those African slaves and their mulatto descendants who once outnumbered whites five to one." Even as she documents a continued Black presence, Thompson concludes that a national politics based on "color" rather than "blood" made for "a vanishing black people: relatively few in numbers, relatively free of racial discrimination and relatively content. Summarized one gentleman, 'If there were more of us, perhaps it would be different.'"[26]

Thompson also relays Josephine Baker's purported exchange in Buenos Aires with Ramón Carillo, the nation's minister of public health. "Where are the Negroes?" Baker asked. His reply: "There are only two. . . . You and I."[27] Harrison was there too. However, Argentina's disavowal of its Black citizens lowers the odds that she called attention to her identity as an African American. Countless Argentineans had silently passed over into whiteness, and she may have had the same impulse, in accord with her credo, "willing to be what ever I can get the best treatments at being" (75). Of course there is no way to know.

Harrison lived in Buenos Aires between 1943 and 1950. Her last known address was 900 Peru Avenue in San Telmo, the city's smallest and oldest district, abutting its southern border. As usual, she situated herself in an alluring and eminently multicultural location. In the seventeenth century San Telmo was packed with brick kilns and warehouses that stored wool, hides, and leather. Over time it became residential, but the yellow fever outbreak of 1871 triggered an exodus of its middle-class residents. Later their vacated homes were converted into tenements for the European immigrants who flooded in. San Telmo had many British, Galician, Russian, Italian, and African residents, with a high rate of intermarriage between them. Now reputedly the city's "más característico" neighborhood, San Telmo's colonial architecture, plazas, churches, and cobblestone streets, along with its museums, galleries, cafes, antique shops, and food halls, have made it a trendy tourist destination. Perhaps Harrison went to mass at San Pedro Telmo, drank café con leche on Plaza Dorrego, and supplemented her instruction in flamenco with tango lessons.

In Cairo, due to her status as a native speaker of English, Harrison was offered governess positions. It is intriguing to consider whether she had the same opportunity in Buenos Aires. However, most likely she mainly supported herself with domestic service work. She would have benefited immensely from the reforms implemented by Juan Perón, which made Argentina a workers' haven in the 1940s. Having campaigned on economic independence, the populist leader sought to implement a five-year plan to achieve full employment, stimulate industrial growth, and increase pay. Workers were guaranteed an annual bonus equivalent to one month's salary—a powerful incentive for Harrison, contrary to her instincts, to bear with a single position for a full year. Perón also arranged housing credits for workers and put price controls on food and other necessities. Between 1945 and 1949, real wages rose by 22 percent.[28]

Of course, this bestselling expatriate American author was no ordinary service worker. In applying for her visa at the Brazilian consulate in San Francisco, Harrison had identified her profession as that of writer—"escritora"—and she continued to actively cultivate that identity in South America. She kept her publishers in New York apprised of her whereabouts and, as we'll see in the epilogue, wrote to American cultural authorities to express her views. She also invested in a typewriter, another sign of how seriously she took her writing.

During the 1940s she was still making some money from her book. Ellery Sedgwick's memoir, *The Happy Profession*, includes an anecdote about assisting her in Argentina: "It was at Buenos Aires that my letters caught up with her. Then she needed my help, for her precious bankbook was on deposit in Boston and the South American Branch was not aware of its existence. A few Vice-Presidents set that right and Juanita moved on to Honolulu."[29] Unfortunately he muddled the chronology. Harrison had not "moved on to Honolulu" from Buenos Aires—she went *from* Honolulu to Brazil. Starting in 1940, she had more than a decade of uninterrupted residency in South America, with no additional Honolulu time sandwiched within it.[30] Yet the germane fact is that all these years later, *My Great, Wide, Beautiful World* still generated royalties. Sedgwick's statement hints, moreover, that they were crucial income. Harrison needed "help" to access the money, and she relied on it to make her next move, to wherever that may have been. It is striking that the famous editor took this service upon himself, sending her multiple notices about the payment she had due and contacting high-ranking bank officials—"Vice-Presidents"—on her behalf. For all his condescension, in this respect he served her well.

Toward the end of Harrison's time in Buenos Aires, the peso began a precipitous slide, losing about 70 percent of its value between 1949 and 1951. Its devaluation sparked massive inflation, corroding workers' purchasing power. Harrison would have seen the signs that it was time to move on, and she was on the road well before salaries hit bottom. She was so often in places at their most economically dynamic moments, and conversely, got out when it started to go south. The turbulence in Argentina mirrored what she had experienced in Havana when the sugar bubble burst, Denver when her bank collapsed, Spain when the US stock market crashed, and France when inflation leaped. Because she always moved toward the opportunities created by societies and economies in flux, it follows that she experienced what might seem like more than her fair share of reversals.

But before leaving South America, first—of course—it was time for a little trip. The impetus for renewing her passport in February 1950 was to travel "by rail" from Buenos Aires to Bolivia, "for pleasure." Far cheaper than Argentina, in Bolivia her savings would have stretched quite far, even as the country's main tourist destinations are hugely scaled. The salt flats at Uyuni were surely on her itinerary, along with perhaps Lake Titicaca—South America's largest lake—and the sprawling ruins of the ancient southern Andes capital, Tiwanaku.

On the passport renewal application, Harrison answered "uncertain" for the question of when she intended "to return to the United States to reside permanently." Yet return was on the horizon. Economically at least, the 1950s were the right time to go back. Between 1948 and 1952, even as Argentina entered recession, the US economy grew by approximately 20 percent. Its strength might have encouraged her to go home, along with an emotional readiness to do so.

While Harrison probably lived the rest of her life in the United States, repatriation did not complete her experience with Latin culture. After all, her next known address was in La Jolla, California. Inspired by popular enthusiasm for *Ramona* and its star-crossed lovers, California chambers of commerce strategized about ways to cash in on the state's Spanish history. Tourists were courted with restored missions and haciendas. Los Angeles built a Hispanic-themed shopping district wholesale. Yet knowing Harrison, along with enjoying such facsimiles, she also connected with actual Latin communities and real Latin people. Steeped in the Spanish language, a sojourner in Caribbean, Spanish, and Southern American cities, a devoted mass-goer and lover of olive oil, flamenco, and bullfights, this American named Juanita savored countless facets of the pan-Hispanic world throughout her life.

CHAPTER 10

Last Years

And now we reach her last years. The final stage of Harrison's life, her mid-sixties and seventies, is not entirely obscure, but for a fourteen-year span I am unsure of her exact whereabouts. I know at least that this period began in La Jolla and ended in Honolulu, and hope in the future to learn more.

Although in 1950 in Argentina, Harrison responded "uncertain" to the question of when she would repatriate, three years later she was in California. As discussed, the factors that induced her to leave Buenos Aires are evident: rising inflation and sinking wages. In addition, after seven years in an inland city, this lover of sea and sun might have found it increasingly dispiriting to live so far from the ocean, in a temperate but rather wet climate.

But what drew her back to the United States? Perhaps she returned almost despite herself, as the path of least resistance for an American citizen. Doing so had many practical advantages. The postwar economy was red-hot, and she could readily plug into the job market. She had a social network in southern California. She was newly eligible to enroll in Social Security. Los Angeles's weather, while inferior to Honolulu's, suited her better than did that of most other places. It is quite possible, moreover, that she anticipated a sense of homecoming in her native country. For all her worldliness, it may have been there that she felt she ultimately belonged.

In the letter to Alice Foster that she wrote from Honolulu's Kapiʻolani Park, Harrison alluded to painful associations with Los Angeles. "I wish nothing from Los Angeles" (FL3), she insisted. At age eighty "the gentle and Kind" George Dickinson had just died, and she mourned him deeply. Moreover, she was put off by what she perceived as a prevailing superficiality. "I did not like the Los Angeles People" (FL2), she had commented while still in Nice. Eventually, though, she did go back—if not to Los Angeles, then not so far from it. 1953 sees her in La Jolla, a community within the municipality of San Diego. For the first time in over a quarter century, Harrison lived in the United States. We can wonder about her experience and emotion adapting to life in that affluent California suburb, after so many years in Argentina's lively capital.

La Jolla is a scant eighteen miles from National City, and Harrison's ties with the Dickinson family may account for why she settled there. As discussed earlier, George's and Myra's families both had roots in National City that reached back to its 1887 incorporation. Given she had been away from the United States for so long, it would have been enormously helpful to have personal connections to get a job.

Harrison had renewed her in-person acquaintance with the Dickinsons in 1937 when several family members vacationed in Honolulu. In respect to her reunion with "Mr and Mrs William George Dickinson," who had visited separately from Myra, she enthused, "The last 10 days have been my happiest."[1] Indicating their continued association, in 1950, a scant three years before she surfaced in La Jolla, she gave Myra's home at 423 South Lafayette Park Place as her legal address, and she specified that "in the event of death or accident," authorities were to "notify Mr. William G. Dickinson, 726 Rives Strong Bldg."[2] The imposing Rives-Strong building was quite the address for an emergency contact—enough of an LA icon to serve as the film location of a window-washing scene in Charlie Chaplin's *The New Janitor*.[3] George and William had moved the company there after selling the Strong & Dickinson building, which was repurposed for the Los Angeles Stock Exchange. We can wonder if the younger Dickinson, following in his parents' footsteps, offered Harrison any financial counsel. Myra survived to the hale age of ninety-six, dying in 1958, and Harrison's late-life residence in southern California could have overlapped with hers by as much as five years. Spanning three generations, the continuity of her Dickinson association encourages speculation over whether she at some point rejoined Myra's household.[4]

Harrison applied for a Social Security card in La Jolla on August 31, 1953, at age sixty-five and a half, just as soon as she was eligible. In respect to the program he signed into law in 1935, President Franklin D. Roosevelt had explained, "We can never insure one hundred percent of the population against one hundred percent of the hazards and vicissitudes of life, but we have tried to frame a law which will give some measure of protection to the average citizen."[5] The payroll taxes that funded worker credits were implemented two years later, with the first monthly benefits paid out in 1940. The transformative program, replacing public almshouses, safeguarded the elderly from indigence. Its Old Age Assist provision was the most significant benefit for the shoulder generation, those who, like Harrison, only began paying into Social Security in their middle age. Harrison eventually received that assistance, and for a woman so frugal, this tiny but guaranteed income must have been a lifeline. As an American citizen back in the United States, she could get a bit of what was due her. In 1937, she had remarked, "The book have sold wonderful but i am still a maid not every day but as often as i need a few cents for my daily bread," adding, "I must keep the book earning for when I am in my 30's cannot be 20 all the time."[6] Her joke may express real anxiety about her dotage. The book royalties did of course peter out, but in time Social Security replaced some of her wage labor income.

In 1953, however, Harrison was still employed, at 7316 Encelia Drive, a four-bedroom house in the La Jolla Country Club Estates subdivision. Her tenure there confirms her enduring ability to get positions in wealthy households. The house was owned by Ruth R. and Arthur T. Heuckendorff.[7] Arthur, a Dane born in Manchuria, was a former "tobacco merchant" for a British company in Shanghai. Perhaps he and Harrison compared their impressions of that city, or of Honolulu, where Heuckendorff had done business shortly after her first sojourn there.[8] The same house was later the residence of Roger C. Guillemin, a French researcher at the Salk Institute who won a Nobel Prize for his brain hormone research. In his biography, Guillemin made a point of describing it, "a Mediterranean house which we have filled, if not overfilled, with many contemporary paintings ... sculptures and potteries," and commemorated "the enjoyable living environment of that happy house."[9] We can hope that while working there, Harrison experienced some of the pleasant atmosphere that Guillemin describes. At the least, she appreciated its sweeping ocean views and proximity to La Jolla Natural Park.

Eventually she left La Jolla for the beach town of Venice, some 130 miles to the north, which was annexed by Los Angeles in 1925. Perhaps the commencement of her Social Security benefits gave her the financial security to relocate. Suggesting she had been somewhat settled there, even after she moved on to Honolulu she retained her bank account with the Venice branch of the Security Pacific National Bank.

The "Venice of America" was founded in 1905 as a resort town by millionaire Abbot Kinney, who sought to elevate a Coney Island–style entertainment complex with classical European architecture. An expanse of marshland was drained into a network of canals, which served to organize the cottage neighborhoods that fanned out from the business block of Italianate arcaded buildings. Who plotted these neighborhoods and brokered their lots, circa 1910? None other than Strong & Dickinson.[10]

Early attractions included a dance hall and amusement pier, a hot salt-water "plunge," and of course the beach. The tourists were joined by a growing number of artists and refugees from Europe. It seems fitting that Venice should have been one of Harrison's homes, given how frequently she dwelled in seaside communities with conspicuous bohemian and European populations, the likes of Havana, Nice, Honolulu, and Mar del Plata. By the time she lived there, Venice's heyday had long since passed: Its canals murky and crime rate cresting, the "Playland of the Pacific" had become the "Slum by the Sea." Yet the fifties were also when Venice began to transition toward a new identity as a funky counterculture community. Harrison was there when the Beats made their presence felt, and their poetry readings and jazz performances at bars and coffeehouses could have sparked memories of what she had experienced in Paris in the 1930s. She was far more constricted spatially, however, than she had ever been in France. Due to restrictive covenants, most of the town's African American residents lived inland in a segregated neighborhood. She was gone before the turbulent events of the late 1960s, during which Venice endured race riots and clashes between the police and the hippies who succeeded the Beats.

Her return to Honolulu was some time after Hawaii became a state in 1959. Prior to statehood, the territory was required to keep immigration records for all arrivals and departures, and 1939 sees the last passenger manifest to bear Harrison's name. That narrows down her arrival time to between 1959 and 1967, probably later rather than earlier. In the 1960s, Honolulu was still small enough that were she in the city for any length of time, one would expect to see some sign of it. After all, she had once hobnobbed with the city's elite, whose doings the local papers minutely reported on.

Harrison was unemployed in Honolulu when she died. Possible evidence that she did not work for wages during that last residence—and its relative brevity—she never opened a local bank account. Instead, she retained the savings account at her Venice bank, keeping just a safe deposit box at the Waikiki Branch of the Bank of Hawaii. (What was in that box, one wonders.) That she managed her finances thus raises the odds that she spent the majority of that uncharted time, from the mid-1950s to the mid-1960s, in California. She may of course have been on the road; knowing Harrison, this should not be ruled out. Yet to readily cash her Old Age Assistance checks, she had to be in the United States. She alluded more than once, moreover, to the prospect of tiring of travel, and as she grew older the likelihood of reaching that saturation point increased.

In 1931, she had commented of her life course, "I have it all set up until I die" (FL2). Was living out her last days in Honolulu part of the "set up"? Sometime in her seventies, she saved up the cost of a ticket, organized her affairs, and gathered her energies to return to this beloved place. Perhaps she was possessed by the same feeling to which Mark Twain confessed: "No other land could so longingly and beseechingly haunt me, sleeping and waking, through more than half a lifetime, as that one has done. Other things leave me, but it abides."[11] Or for more contemporary nostalgia, from the forty-fourth president of the United States: "Even now . . . I can retrace the first steps I took as a child and be stunned by the beauty of the islands. The trembling blue plane of the Pacific. The moss-covered cliffs and the cool rush of Manoa Falls, with its ginger blossoms and high canopies filled with the sound of invisible birds. The North Shore's thunderous waves, crumbling as if in a slow-motion reel. The shadows off Pali's peaks; the sultry, scented air."[12] Unlike Twain, Harrison was able to return to that haunting place in the flesh, to the Hawaii that Barack Obama conjures in *Dreams from My Father* in describing the sights, sounds, and scents of his childhood, which coincided with her last Honolulu residency.

She didn't exactly score the "little future nest in Hawaii" (311) that she had once envisioned, if we interpret that to be private homeownership. At the time of her death, Harrison was living in a rental unit in Aloha Suites on Lemon Road, a modest apartment building of the kind that by mid-century was ubiquitous in Honolulu. Aloha Suites is next door to one of the city's early wooden bungalows, which might have caused her to reminisce about the Tadas' cottage on 105 Ohua Avenue. Coincidentally, or not, it is also kitty-corner from the hostel where I was staying when I discovered she had lived there—I could see her apartment from my dorm room. That same hostel had been converted from the apartment complex where the Tadas lived after the attack on Pearl Harbor.

The profusion of apartment buildings is only one of many signs of the transformation Honolulu underwent during the decades Harrison was away. The traumatic years of martial law had long since passed, and statehood launched an extended period of economic growth. Between 1958 and 1973, per capita personal income increased at an annual rate of 4 percent.[13] Yet the economy was still in transition. Mass tourism, although galvanized by the regular jet service that commenced with statehood, was still not quite Hawaii's mainstay. Pineapple, sugar, and now the military were the primary drivers. As late as the 1970s, hotels on Oahu only numbered in the dozens, and the historical renovation that gave the Royal Hawaiian and Ali Moana hotels a second lease on life had not yet begun. The run-down condition of these flagship properties must have dismayed Harrison, who had once bragged about them to her friends.

On the social front, Hawaii's movement toward intercultural solidarity was advancing, and in accord with the civil rights movements taking place across the nation, it saw an outpouring of Indigenous activism. Its multiethnic society was seen as an exemplar by those fighting for equal rights in the South. Addressing the new legislature, Martin Luther King Jr. stated, "We look to you for inspiration and as a noble example, where you have already accomplished in the area of racial harmony and racial justice what we are struggling to accomplish in other sections of the country."[14]

As for Harrison's private circle, Kuni Tada was still in Honolulu, outliving her by some years. Perhaps the two women resumed their friendship, and Harrison associated with some of the younger Tadas upon whom she had once "lavished" gifts."[15] Hopefully her monthly government checks, coupled with her thrift, enabled her to pass in retirement the leisurely days she cherished. We can imagine her life there—swimming in the sea, dining at outdoor cafes, socializing with all and sundry.

Harrison's death certificate and probate documents are on microfilm at the Honolulu courthouse. Some of their revelations include how much money she had in her bank account when she died ($401.04), marital status (single, always single), last address (2569 Lemon Road), cause of death (cancer), and burial at Valley of the Temples Memorial Park.

The park is still another of the storied places that Harrison habitually occupied, now even in death. Valley of the Temples has become an upscale facility, encompassing 240 acres. Ferdinand Marcos's remains were held there while his widow, Imelda Marcos, sought permission to repatriate them to the Philippines—she is rumored to have defaulted on the tremendous electricity bill for his refrigerated crypt. Yet on its opening in the mid-1960s, Valley of the Temples was a humble place. Brand new and on the windward side of Oahu, it was much cheaper than the older urban cemeteries, and its target customers were plantation workers. As it expanded, it developed around ethnic "neighborhoods," sections dedicated to Chinese, Japanese, Filipino, and Anglo interment.

My journey from Waikiki to Valley of the Temples to visit Harrison's grave took close to three hours on local buses, over the steep Pali pass and then north through the valley at the base of the jaw-dropping Ko'olau Range. Since I knew the plot number, I had thought I could just wander the grounds until I found it. This huge complex, however, was not the kind of graveyard that one ambles through, and the security guard at the front gate directed me to the funeral home. An outdoor memorial service was underway, bringing home to me that I was at a real cemetery.

On calling up Harrison's record, staff members found that she was in the farthest reach of the place. One of them, Pat Lau, offered to drive me there, and along the way, she gave me a tour. Her previous job had in fact been that of tour guide, and it was a valuable history lesson, with occasional notes of sales pitch—as when she extolled the facility's "hedge estates," family plots surrounded by a shin-high row of bushes. She pointed out the different ethnic areas and their special features, such as the Japanese Buddhist temple and meditation pavilion. She also stated, persuasively, that one could feel the power of the mountains that loomed at its western border.

But we drove on, leaving the lush areas behind. Harrison is buried at the edge of the complex, in the first area to receive remains when it opened. Overlooking the busy main road, the section is a stretch of rough lawn abutting the rainforest. It is maintained but not landscaped.

The majority of the sites bear a brick-sized marker, inscribed with only name and lifespan: Juanita Harrison, 1887–1967. With some hesitation, Lau told me that such a marker indicated the deceased was a "ward of the state." While such status did not necessarily connote indigence, it did point to an "unattended death," meaning that the government had stepped in to arrange burial. It was startling to hear her repeatedly refer to Harrison as a ward. My research trail had possibly ended at an actual pauper's grave.

Before leaving me there for reflection, Lau gave me one of the spare bento that the mourning family had donated to the staff after their service. I was hungry and ate all that rice, fish, and chicken too fast, finding myself on the verge of choking on funeral meats. It flashed through my mind that were I to be found expired, lying by Harrison's grave, the conclusion might be that I was overcome by learning about the nature of her final days.

Certainly her not having the friends and money to arrange the interment and cover its costs could seem sad. A poignant detail is that the mortuary collected $12.66 from her estate to purchase clothing for her burial. Her fate might recollect Zora Neale Hurston's resting place at the Garden of Heavenly Rest in Fort Pierce, Florida, as memorialized by Alice Walker in her 1975 essay, "In Search of Zora Neale Hurston."[16] From the funeral home that received Hurston, Walker learned that "folks took up a collection" for her burial expenses. Any records had long since vanished, and to find her grave Walker searched a fearsome expanse of waist-high weeds until at last she stumbled into a six-by-three-foot sunken rectangle that she concluded must be it. Hurston had spent her last days at the St. Lucie County Welfare Home, and prior to that she supported herself with occasional housekeeping jobs, just as Harrison did once her book royalties fell off.

Nonetheless we are not obliged to read Harrison's end as tragic (nor Hurston's, of course). On the contrary, it can be conceived as triumphant. At the time of her death, just short of her eightieth birthday, she was living independently, on a spectacular island that she had deliberately chosen as her final home. She had no close friends to attend to her burial—but then, she was wont to complain, friends could be such a drag. Unlike Hurston, moreover, she was laid to rest in not a segregated facility but one that broadcasts its pluralism, amid an assembly of Filipinos, Chinese, Japanese, Hawaiians, Germans, and Anglos. This section of the park predates its organization into ethnic subdivisions, and the interred to the right of her include Pedro Queria—her immediate neighbor—Margaret L. Ambrose, Victoriano Ubay, and Taki Kurushima. The names on the surrounding markers are equally diverse: Felix Borja, Rachel L. Krempa, Victoriano Kilip, Philip Kapeliela, Charles K. Wilson, Francisco Cabogo, Mong Ho, Donald Jacobs, Rachel Batabalona, Thomas Ng, Inoruto T. Mosion, Irene K. Haumea, Clarence Pi Jr., Melicio Peru, and the infants Jesse Jensen and Mary M. Jumawa. Such multicultural community was just what she liked.

We can in a like manner reevaluate her small estate, as not pitiful but fitting. To preserve her freedom, throughout her life Harrison insisted on shedding encumbrances. Proud to carry the bare minimum of luggage that proved her to be "the best Traveler" (59), she valued not material goods but other kinds of possessions. Her dedicated roaming, the title of her book insists, made the entire world hers.

That said, she was hardly indigent. She had an income and money in the bank, just little to spare. Harrison once declared, "I do not need to save for cold days" (311). Her California real estate investments during the heady mid-1920s aside, she did not rely on capitalist accumulation. Her fearless habit was to cyclically save and spend. Given that she budgeted her trips so as to arrive at her next resting point with no surplus funds, she surely would have seen it as wasteful miscalculation to take money to the grave. And why pay the full costs for one's burial when the government could help out?

In 1942, *My Great, Wide, Beautiful World* was listed in the Honolulu Public Library's special collection of books by African American authors. It had illustrious company, volumes of fiction and nonfiction by major Harlem Renaissance and midcentury writers that include Marian Anderson, Countee Cullen, Jessie

Fauset, Langston Hughes, James Weldon Johnson, and Richard Wright. *Honolulu Advertiser* suggested the collection could offer a feeling of homecoming in encouraging "Negro troops from Harlem" to "take time off from their shooting practice occasionally and visit the library."[17] Harrison was literally cataloged as African American on the island where she was once extolled as "Part American Negro, Part American Indian and Part just plain American."[18]

Yet at her death twenty-five years later in the same city, she was identified as white. I almost missed learning this. I'd woken with a headache on the last day of my research trip, slated for squeezing in a visit to the courthouse before my late-night flight, and by the end of the stop-and-start bus ride downtown it had bloomed into a migraine. Scrolling through the microfilm of her probate records didn't help. Once I got them printed, I went to the women's room to throw up. Waiting for the next wave in the stall, trying to shut out the sounds to the right and left of me, I passed the time reviewing the papers, only to realize that portions of the death certificate were illegible. After returning to the archive to have it reprinted, I saw that it identified her race as "Caucasian."

The designation might have been due to claims that Harrison herself made, in person or on paper. It might have been the conclusion of a hospital or social worker—a doctor, nurse, or administrator. That she was deemed white could also indicate her distance from a local African American community, a dearth of Black visitors.

Looking beyond her individual circumstances, Hawaii's unique demographics created the opportunity for the misconception. Whereas in the Mississippi of Harrison's youth, the least quotient of Black ancestry resulted in one being identified as Black, racial categories in Hawaii were far more various and flexible. The majority of its denizens were of mixed heritage. Yet ironically enough, Harrison was assigned a default whiteness, baldly deemed Caucasian in this racially fluid state.

Perhaps in her last years she did live as white, enabled by her lapse into obscurity. Three decades after the publication of her only book, she was no longer a celebrated African American author. One might be tempted to present this as the end of her racial story, a neat narrative arc from her birth as a Black child in Jim Crow Mississippi to her death as a white woman in the freshly minted state of Hawaii. Yet all I have learned about Harrison makes it hard to imagine her as an "ex-colored" woman, someone like James Weldon Johnson's antihero, who became "an ordinarily successful white man" suspecting he had surrendered his "birthright."[19] More likely is that she engaged in a form of disidentification, keeping some distance from a coalitional racial identity, even as her Blackness—"her cosmopolitan brown complexion," to use Chimene Jackson's phrase—helped her make connections across national and ethnic lines.[20] A continuation of her lifelong methods, far from a rejection of her heritage, positioning herself thus fostered her identity as a citizen of the world.

My Great, Wide, Beautiful World, as we know, dramatizes this practice, expressing her complex subjectivity far more vividly than could any scholarly pontification. This book opened with Harrison's encounter with a group of African women at the Jardin d'Acclimatation, of whom she observes, "They took a fancy to me. I think they saw I had some of their blood I couldnt fool them" (19). In this and many of the other situations she describes, it is unclear just where her "fooling" might reside—in deliberate acts and statements or diffused across her overall behavior and self-presentation. Overseas, she was displaced from not only her community of origin but also larger American social contexts. This distance opened her ethnicity and nationality to endless interpretation, which she recorded with zeal. "If I go out without a Hat," she notes in Naples, "the Italians do not take any notice of me and always talk right along with me But if I have on a Hat they call me a Chinese or Japenese" (38). In Djibouti, despite her donning a "black Vail head dress" (160), Somali women assessed her as Greek or Chinese. While living in the south of France, to dodge Yankee stigma she claimed to be Cuban. The Spanish dialect she had learned in Cuba caused Spaniards to wonder if she were an Argentine, while customs officials in Japan were confounded by the American passport that this seemingly impecunious woman carried. And in Budapest: "They think I am Italian and am makeing believe when I say I am American I just leave it to them" (53).

From Jerusalem, she likewise recounts, "I have a very Oriental looking scarf I ware most of the time on my head everyone think I am Arabian but are puzzled to see me with such a short french dress and the first thing they ask my Friend If I am Arabian then when I ware my little French cap they take me for Jewish." She concludes the anecdote by stating, "I am willing to be what ever I can get the best treatments at being" (75). Sometimes she directed how she was perceived; sometimes others drew their own conclusions according to her dress, language, appearance, and social and actual locations. Being judged Caucasian at her Honolulu deathbed was more of the same. Was it a category she appropriated or one that was foisted on her? Or a bit—somehow—of both? Regardless, she surely would have been amused at being officially classified as white, the final entry in her life ledger of attribution.

EPILOGUE

"A True Literature of American Life"

In accord with its title, this study has focused on Harrison as a writer, but of course she was a reader, too. *My Great, Wide, Beautiful World* shows her perusing travel guides, haunting libraries, and dedicating her last pennies, or rupees, to books. No passive reader, she confidently assessed which authors were "the best" (118). Her response to a 1913 lecture on *Paradise Lost* at Carnegie Hall proves an early readiness to enter into critical debate: "I do not agree and have ofen felt sorry for the speaker" (26). She was daunted neither by the illustrious venue and authoritative speaker nor by the highbrow poem itself.

Some thirty years later, she brought this striking interpretive assurance to the mid-1940s letter she wrote to theater critic John Mason Brown, who had a regular column in *Saturday Review of Literature*. Brown had argued that performances of *Uncle Tom's Cabin* should be encouraged, not shunned. Although Harriet Beecher Stowe's 1852 novel galvanized the abolitionist movement, its racial stereotypes are flagrant. The enslaved people to whom Stowe grants the fullest humanity, the greatest intelligence and refinement, are the lightest skinned. The darker characters are more usually comic foils or, in Tom's case, a sacrifice. Tom does not fight for his freedom or even his life. Sam and Andy are straight out of minstrel shows, Topsy is figured as an imp, and Stowe mostly forgets about the woman who manages "Uncle Tom's Cabin"—Chloe, Tom's bereft wife—only occasionally checking in on her anger and grief.

Protestors in Bridgeport, Connecticut, had objected that *Uncle Tom's Cabin* "refreshed memories that tend to portray only the weaknesses of a racial minority."[1] Addressing the controversy, Brown maintained, "One would have thought that if any book merited the gratitude of those interested in racial equality, it would have been that volume."[2] Writing to Brown to second—at least "some what"—his defense of *Uncle Tom's Cabin*, Harrison went well beyond simply agreeing with him to decry the erasure of African Americans from the cultural landscape. Here is that letter in full—original punctuation, misspellings, ellipses, and typos all preserved:

Mr. Mason Brown

Mr. Brown In your finding no fault in the showing of uncle tom's cabin on town meeting, May 9. I have to agree with you some what as Negro are gagging over the fly and holding an elephant in the mouth, the elephant's trunk in the eyes and that cloud the thinking, If the time wested objecting to the showing of uncle tom's cabin was put in to haveing themselves included in the textbooks of the American schools, and in the discrispion of that man God made in his own image.

The Negro let the white man of America ex culde him from everything worth while, textbooks all holy descrispion of God's creation. . . . I think if the Negro would object to being excluded from the part of life that go to make for love, and not HATE. Then there May be room for that song (BROTHERHOOD) and a true literture of American life, that would bring out the best that is in a people; and not just the clowing part of some. . . . Why are the white people so carefull not to include the Negro in the holy Literture and their textbooks? . . Mr. Brown, If the truth were told and had been told all along there would be no need for uncle tom's.

Why no church gives any account of the black people on earth? I have a Catholic textbook, that is use in Negro schools it give gives the child an account of every thing the child has to live with, and see in all his life, but the black people. why?? The Negro is surely not included in the christian Literture.

As a Catholic myself, I know the sepration of the races is one of the stronges point of the Catholic Church. From childhood I sew it . . . I am wondering if God know there are black people on earth? . . Or, are they to turn white before God will addmitte them. you want to show uncle tom's cabin for the sake of Amerian culture. Why not show some part off uncle tom in your American

Chrisrian Literture, and textbooks for the sake of culture? It may help the Negro, if no one eles?

Too bad the Negro has adopted a religion that he is not Identify with outwardly...... I belive that is why we are the foot... stool, of every one. Becouse we serve every thing and every body, but God himself.

She signed herself "Mrs. Juanita Harrison."[3]

There is so much one could say about this passionate, roughly executed letter. The place to start is that Harrison was a horrible typist. Between the mangled typescript and the elephant and fly metaphor, on discovering the letter at Harvard University's Houghton Library, for some moments I could make no sense of its opening lines. Yet the very fact that she was typing indicates how seriously she continued to take her writing.

The letter helps us track her practice during those thinly documented years in South America, showing her commitment to upholding her professional identity. She likely wrote it in late 1945 or 1946 while living in Buenos Aires, since the *Saturday Review* column to which she responded ran in the fall of 1945. The letter proves that she kept up with American periodicals from abroad and retained the habit of international correspondence, mailing off dispatches to the United States. One wonders to what other public figures Harrison wrote and what other stray missives are still to be found.

While composing the letter, perhaps she thought back to criticism of her own book. Despite the widespread praise for *My Great, Wide, Beautiful World*, some readers were troubled by its comic aspects, "likely to thrill," in Arthur Spingarn's words, "only those people who know 'colored people are like that.'"[4] Harrison herself, however, was unconcerned as to whether her narrative might be culled for racial stereotypes, whether it sometimes expressed, as she puts it here, "the clowing part." The letter demonstrates the ease with which she could assume the role of cultural spokesperson, with the authority to weigh in on debates. The world that she claimed as "my" encompassed the realm of print culture. She was enough of a personage, moreover, for Brown and his executors to safeguard her letter. That it should surface in the rarefied environment of Harvard's special collections in Cambridge speaks to her paradoxically scaled life. Despite her itinerant habits and sometimes precarious lifestyle, she had a way of being at cultural hubs and of associating with influential people, who treated her with respect.

The letter matters most for the light it sheds on her racial views. Harrison confidently entered a debate that ranged beyond *Uncle Tom's Cabin* to point out pervasive problems with African American representation in US discourse at large. It was crucial to "a true Literture of American life" that religious and educational texts should attest to Black presence. She is explicit about the conditions that account for "why we are the foot ... stool, of every one," explicit about what Black Americans and white Americans must do to effect change.

Wielding a coalitional "we," this letter makes for very different reading than *My Great, Wide, Beautiful World*, in which Harrison celebrates her ability to assume a multitude of identities, and in which racial matters are usually implicit. Her fierceness here makes one realize how deliberate she must have been in the letters that comprise the book, in choosing what to repress. The difference reminds us, once again, of the book's original rhetorical form—of the first audience and initial purpose of those long travel accounts—as she managed a set of crucial relationships to support the experiences she desired.

Finally, more personally, thinking about *Uncle Tom's Cabin* prompts memories of her Southern past. There is a telling slippage, or elision, in her statement, "As a Catholic myself, I know the sepration of the races is one of the stronges point of the Catholic Church. From childhood I sew it." At first reading, this sentence seems to indicate that "from childhood," Harrison "saw" racial segregation as practiced by the Catholic Church. However, she only became a Catholic in adulthood and by all evidence had little exposure to that religion prior to her travels. "The sepration of the races" that she observed and experienced as a girl refers more generally to the segregation endemic to the late nineteenth-century South. It makes sense that such memories should surface, and that a collective identity should emerge, in discussing a novel that showcases her region of origins. *Uncle Tom's Cabin* maps the terrain, culture, and laws of the South. Even though Harrison does not explicitly reference her native Mississippi, the novel transports her there. I am astonished that it should be *this* book that the letter discusses, to date the latest extant text in her hand.

Chimene Jackson has characterized Harrison's achievement as healing "cultural wounds" through "owning herself after years of Blacks being owned; her reversal of the transatlantic portion of the Middle Passage; her uncontested vehemence, sass and survival during decades when it was forbidden for Blacks to look a white in the eye; her sleeping both outside and in hotels and her intellectual engagement with whites—scorning classism and white supremacy in a few languages. And finally, her ability to choose her final destination, always."[5]

Writing and publishing a book about these experiences further closed those wounds. Her editor named her a "born writer" as they worked together on the manuscript in Paris. Although she did not utter it, Harrison preserved in writing her silent, private response to the compliment: "I thought how far I was from what I was born" (FL2). The vast distance she traveled over the course of her life can be measured in not only miles but also achievement: from her impoverished, disenfranchised Mississippi childhood to multilingualism, authorship, celebrity, and world citizenship, a confident actor in all the realms of her "great, wide, beautiful world."

Notes

Introduction

1 Juanita Harrison, *My Great, Wide, Beautiful World*, with a preface by Mildred Morris (Macmillan, 1936); reprint, G.K. Hall & Co., 1996, with an introduction by Adele Logan Alexander. Hereafter parenthetically cited with just a page number.
2 I use African American and Black interchangeably in referring to Harrison's racial identity. I also include, without comment on their terminology, quotations from contemporary sources that identify her as colored or Negro.
3 William Norwood, *Honolulu Star-Bulletin*, March 25, 1936, p. 4.
4 Korey Garibaldi, *Impermanent Blackness: The Making and Unmaking of Interracial Literary Culture in Modern America* (Princeton University Press, 2023), 91.
5 Michael Ra-Shon Hall, *Freedom Beyond Confinement: Travel and Imagination in African-American Cultural History and Letters* (Clemson University Press, 2021), 3.
6 Virginia Whatley Smith, "African American Travel Literature," in *The Cambridge Companion to American Travel Writing*, ed. Alfred Bendixen and Judith Hamera (Cambridge University Press, 2009), 197–213.
7 Elizabeth Stordeur Pryor, *Colored Travelers: Mobility and the Fight for Citizenship Before the Civil War* (University of North Carolina Press, 2016), 8.
8 Qtd. in Gary Totten, *African American Travel Narratives from Abroad: Mobility and Cultural Work in the Age of Jim Crow* (University of Massachusetts Press, 2015), 96.
9 Pat Noxolo, "Geographies of Race and Ethnicity 1," *Progress in Human Geography* 46, no. 5 (2022), 1232–40.
10 Totten, 5, 6.
11 Chimene Jackson, "Black Women's Journals Reflect Mine, Yours, and Ours: Through the Travel Writing of Juanita Harrison," in *Diary as Literature: Through the Lens of Multiculturalism in America*, ed. Angela R. Hooks (Vernon Press, 2020), 76.
12 *Publisher's Weekly*, April 25, 1936, p. 1670.
13 Zalka Csenge Virág, March 14, 2018, http://multicoloreddiary.blogspot.com/2018/03; Kathy, July 2010, https://www.goodreads.com/book/show/711127.
14 Juanita Harrison to Alice Foster, letter, March 11, 1931, Ann Cunningham Smith collection of letters from Juanita Harrison to Alice M. Foster, Charles E. Young Research Library, University of California, Los Angeles. Hereafter parenthetically cited as FL2.
15 Kali Nicole Gross, *Hannah Mary Tabbs and the Disembodied Torso* (Oxford University Press, 2018), 4.

16 Gross, *Hannah*, 5.
17 Kali Nicole Gross, "'Ordinary Yet Infamous': *Hannah Mary Tabbs and the Disembodied Torso*," *Not Even Past*, February 1, 2016.
18 Era Bell Thompson, *American Daughter* (University of Chicago Press, 1946), 288.
19 Marlon Ross, *Manning the Race: Reforming Black Men in the Jim Crow Era* (New York University Press, 2004), 35.
20 Adam Bledsoe, Latoya E. Eaves, and Brian Williams, "Introduction: Black Geographies in and of the United States South," *Southeastern Geographer* 57, no. 1 (2017), 9.
21 Katherine McKittrick, *Demonic Grounds: Black Women and the Cartographies of Struggle* (University of Minnesota Press, 2006), xii.
22 Judith Madera, *Black Atlas: Geography and Flow in Nineteenth-Century African American Literature* (Duke University Press, 2015), 5.
23 Madera, *Black*, 1.
24 Madera, *Black*, 3.
25 Madera, *Black*, 1.
26 Carolyn Steedman, *Dust: The Archive and Cultural History* (Rutgers University Press, 2002), 8; Eric Gardner, "Sowing and Reaping: A 'New' Chapter from Frances Ellen Watkins Harper's Second Novel," *Common-Place: The Interactive Journal of Early American Life* 13, no. 1 (October 2012); John Kevin Young, *Black Writers, White Publishers: Marketplace Politics in Twentieth-Century African American Literature* (University Press of Mississippi, 2006), 22; Judith Madera, "Shaking the Basemap," in *The Black Geographic: Praxis, Resistance, Futurity*, ed. Camilla Hawthorne and Jovan Scott Lewis (Duke University Press, 2023), 50; Lois Brown, "Death-Defying Testimony: Women's Private Lives and the Politics of Public Documents," *Legacy* 27, no. 1 (2010), 138.
27 Respectively, in *Maverick Autobiographies: Women Writers and the American West, 1900–1936* (University of Wisconsin Press, 2004), *Playing House in the American West: Western Women's Life Narratives, 1839–1987* (University of Alabama Press, 2013), and *Faraway Women and the "Atlantic Monthly"* (University of Massachusetts Press, 2019).
28 Jill Lepore, "Historians Who Love Too Much: Reflections on Microhistory and Biography," *Journal of American History* 88, no. 1 (June 2001), 135.
29 Jordan Alexander Stein, review, *The Notorious Elizabeth Tuttle: Marriage, Murder, and Madness in the Family of Jonathan Edwards*, by Ava Chamberlain, *Early American Literature* 48, no. 3 (2013), 799.
30 Saidiya Hartman, "Venus in Two Acts," *small axe* 26 (June 2008), 12; *Wayward Lives, Beautiful Experiments: Intimate Histories of Riotous Black Girls, Troublesome Women, and Queer Radicals* (W.W. Norton & Co, 2019), xiv.
31 Ayesha K. Hardison, *Writing Through Jane Crow: Race and Gender Politics in African American Literature* (University of Virginia Press, 2014), 3.
32 Qtd. in Martha Cutter, "Why Passing Is (Still) Not Passé After More Than 250 Years: Sources Past and Present," in *Neo-Passing: Performing Identity After Jim Crow*, ed. Mollie Godfrey and Vershawn Ashanti Young (University of Illinois Press, 2018), 54.
33 Lloyd E. Ambrosius, introduction to *Writing Biography: Historians and Their Craft* (University of Nebraska Press, 2004), ix.

34 Hortense Spillers, "'Mama's Baby, Papa's Maybe': An American Grammar Book," *Diacritics* 17, no. 2 (1987), 65.
35 Leon Jackson, "The Talking Book and the Talking Book Historian: African American Cultures of Print: The State of the Discipline," *Book History* 13 (2010), 269.
36 See Godfrey and Young, *Neo-Passing*.
37 I borrow the term "hostile geographies" from Stacie McCormick, "We Are Here: Jesmyn Ward's Black Feminist Poetics of Place in *Men We Reaped*," *a/b: Auto/Biography Studies* 38, no. 2 (2023), 543–58.
38 In one of her letters, Harrison signs her name with the middle initial "V." In theory, that initial should distinguish her from the other Juanita Harrisons who came of age in the early twentieth century, but this is the single instance I have seen. It is enjoyable though to speculate on what the "V" might have stood for. In 1887 very few American girls' names started with the letter. The most common was Viola.

Chapter 1

1 Eileen Whitfield, *Pickford, the Woman Who Made Hollywood* (University Press of Kentucky, 1997), 246.
2 Frederick Douglass, *My Bondage and My Freedom* (Miller, Orton & Mulligan, 1855), 34.
3 Qtd. in Judith Madera, "Shaking the Basemap," in *The Black Geographic: Praxis, Resistance, Futurity*, ed. Camilla Hawthorne and Jovan Scott Lewis (Duke University Press, 2023), 52.
4 Juanita Harrison to Alice M. Foster, letter, July 12, 1936, Ann Cunningham Smith Collection. Hereafter parenthetically cited as FL3.
5 Patricia Yaeger, *Dirt and Desire: Reconstructing Southern Women's Writing* (University of Chicago Press, 2000), 59.
6 Respectively, on Harrison's 1940 visa for Brazil (Creglar), 1950 passport renewal application at the US consulate in Buenos Aires (Crutler), and 1953 application for a Social Security account number (Creagler).
7 US Census, 1870, Oktibbeha, Mississippi. https://www.familysearch.org/ark:/61903/1:1:MFSL-T55. Entry for Lizzie Crigler.
8 James Cole, *Oktibbeha County* (Arcadia Publishing, 2000), 30.
9 Cole, 7.
10 "Largest Slaveholders from 1860 Slave Census Schedules and Surname Matches for African Americans on 1870 Census," Oktibbeha County, Mississippi, transcription Tom Blake, 2001, http://files.usgwarchives.net/ga/crawford/census/1860/slave.txt.
11 "Largest Slaveholders."
12 "Sandfield Cemetery," Mississippi Civil Rights Project, June 2020, https://mscivilrightsproject.org.
13 Barbara Krauthamer, *Black Slaves, Indian Masters: Slavery, Emancipation, and Citizenship in the Native American South* (University of North Carolina Press, 2013), 41.
14 Greg O'Brien, "Treaty of Dancing Rabbit Creek (1830)," *Encyclopedia of Alabama*, http://encyclopediaofalabama.org/article/h-3426.
15 Obituary, John Lewis Crigler, *East Mississippi Times*, August 30, 1912, p. 1.
16 Raquel Thiebes, "Tracing Your Ancestors in Sturgis Oktibbeha County," November 22, 2005, http://sturgisms.homestead.com/slave.html.

17 Korey Garibaldi, *Impermanent Blackness: The Making and Unmaking of Interracial Literary Culture in Modern America* (Princeton University Press, 2023), 80.
18 Frederick Parks, Angela Crigler, "Black Criglers," April 20, 2001, https://www.ancestry.com/boards/surnames.crigler/40.41.46.49.
19 United States Census (Slave Schedule), 1860, Oktibbeha, Mississippi. FamilySearch, https://www.familysearch.org/ark:/61903/1:1:WKLR-LPT2. Entry for Green B Stallings.
20 Krauthamer, 37.
21 Krauthamer, 41.
22 Krauthamer, 44, 42.
23 Samuel J. Wells, "The Role of Mixed-Bloods in Mississippi Choctaw History," *After Removal: The Choctaw in Mississippi*, ed. Samuel J. Wells and Roseanna Tubby (University Press of Mississippi, 1986), 51.
24 Wells, 51.
25 Wells, 46; Dennis J. Mitchell, *A New History of Mississippi* (University Press of Mississippi, 2014).
26 Mitchell.
27 Wells, 52.
28 Krauthamer, 43, 44.
29 Digitized Index to the Final Rolls of Citizens and Freedmen of the Five Civilized Tribes in Indian Territory (Dawes), 140, 60, https://www.archives.gov/research/native-americans/rolls/final-rolls.html.
30 *Battle Creek Enquirer*, July 20, 1937, p. 8; James T. Hamada, "Juanita Harrison, *Atlantic Monthly* Writer Now Here, Is Mighty Fond of Japanese Baby," *Nippu Jiji*, April 22, 1936, p. 1; William Norwood, *Honolulu Star-Bulletin*, March 25, 1936, p. 4.
31 Zora Neale Hurston, "How It Feels to Be Colored Me," *The World Tomorrow* 11 (1928), 215–16.
32 Chimene Jackson, "Black Women's Journals Reflect Mine, Yours, and Ours: Through the Travel Writing of Juanita Harrison," in *Diary as Literature: Through the Lens of Multiculturalism in America*, ed. Angela R. Hooks (Vernon Press, 2020), 83.
33 Qtd. in Judith Madera, *Black Atlas: Geography and Flow in Nineteenth-Century African American Literature* (Duke University Press, 2015), 12.
34 Jennifer Ritterhouse, "Daily Life in the Jim Crow South, 1900–1945," *Oxford Encyclopedia*, May 24, 2018, https://doi.org/10.1093/acrefore/9780199329175.013.329.
35 Stephanie M. H. Camp, *Closer to Freedom: Enslaved Women and Everyday Resistance in the Plantation South* (University of North Carolina Press, 2004), 9.
36 Isabel Wilkerson, *The Warmth of Other Suns: The Epic Story of America's Great Migration* (Random House, 2010), 31.
37 "Sandfield."
38 Jennifer Baughn, MDAH, "White Arches" (Harris-Banks House), *Historic Resources Inventory*, June 2020, https://www.apps.mdah.ms.gov; Obituary, Lucy Harrison Duncan, *The Meridian Evening Star*, September 29, 1902, p. 6.
39 Leon Jackson, "The Talking Book and the Talking Book Historian: African American Cultures of Print—The State of the Discipline," *Book History* 13 (2010), 267.
40 David H. Jackson Jr., "Segregation," *Mississippi Encyclopedia*, July 11, 2017, http://mississippiencyclopedia.org/entries/segregation/.

41 Bridget Smith Pieschel, "The History of Mississippi University for Women," *Mississippi History Now* (Mississippi Historical Society, March 2012), n.p.
42 David M. Katzman, *Seven Days a Week: Women and Domestic Service in Industrializing America* (University of Illinois Press, 1981), 216–17.
43 Ayesha K. Hardison, *Writing Through Jane Crow: Race and Gender Politics in African American Literature* (University of Virginia Press, 2014), 15.
44 Katzman, 193.
45 Linda K. Kerber, *No Constitutional Right to Be Ladies: Women and the Obligations of Citizenship* (Farrar, Straus and Giroux, 1999), 70.
46 "Railroad Wreck with One Hundred Victims," *New York Times*, September 4, 1902, p. 1.
47 Yaeger, 60.
48 "18 die in Train Wreck in Czechoslovakia," *New York Times*, September 11, 1928.
49 Katzman, 200.
50 Katzman, 197.
51 Katzman, 196.
52 Katzman, 197.
53 "More Slavery at the South, by a Negro Nurse," *The Independent* (January 25, 1912), 196–200, https://docsouth.unc.edu/fpn/negnurse/negnurse.html. Qtd. in Rhonda Bellamy and Bertha Boykin Todd, *Meet the Help: True Stories of Domestics* (Dorrance Publishing Company, 2015), 48.
54 *Meet the Help*, 49.
55 Katzman, 204, 208.

Chapter 2

1 "Juanita Harrison to Be Introduced," *Honolulu Star Advertiser*, August 12, 1937, p. 5.
2 Lawrence R. Rodgers, "Dorothy West's *The Living Is Easy* and the Ideal of Southern Folk Community," 167, *African American Review* 26, no. 1 (1992), 161–72.
3 Carole Marks, "Black Workers and the Great Migration North," 148, *Phylon* 46, no. 2 (2nd Qtr., 1985), 148–61.
4 James Weldon Johnson, *Black Manhattan* (Alfred A. Knopf, 1930); reprint (Atheneum, 1975), 151.
5 Carole Marks, *Farewell, We're Good and Gone: The Great Black Migration* (Indiana University Press, 1989), 94.
6 Saidiya Hartman, *Wayward Lives, Beautiful Experiments: Intimate Histories of Riotous Black Girls, Troublesome Women, and Queer Radicals* (W.W. Norton & Co., 2019), 108, 45–46.
7 Qtd. in Emily Lutenski, *West of Harlem: African American Writers and the Borderlands* (University Press of Kansas, 2015), 107.
8 Susan Armitage, "Western Women's Biographies," *Western American Literature* 41, no. 1 (Spring 2006), 71.
9 Richard Wright, *Black Boy: A Record of Childhood and Youth* (Harper Perennial Modern Classics, 2006), 415.
10 Juanita Harrison, passport application, US consulate Buenos Aires, February 17, 1950. Private collection.

11 Joe William Trotter Jr., *The Great Migration in Historical Perspective: New Dimensions of Race, Class, and Gender* (Indiana University Press, 1991), xi.
12 R. H. Leavell, qtd. in Trotter, 6.
13 Charles S. Johnson, "Chicago Study, Migration Interviews," 1917, qtd. in "'We Tho[ugh]t State Street Would Be Heaven Itself': Black Migrants Speak Out," *History Matters*, http://historymatters.gmu.edu/d/5337/.
14 James Grossman, *Land of Hope: Chicago, Black Southerners, and the Great Migration* (University of Chicago Press, 1991), 7.
15 Grossman, 4.
16 Grossman, 4.
17 Lawrence R. Rodgers, *Canaan Bound: The African-American Great Migration Novel* (University of Illinois Press, 1997), 70.
18 Grossman, 4.
19 Morgan Jerkins, *Wandering in Strange Lands: A Daughter of the Great Migration Reclaims Her Roots* (HarperCollins, 2020), 12.
20 Langston Hughes, *One-Way Ticket* (Alfred A. Knopf, 1949), 61.
21 Nancy Beth Jackson, "If You're Thinking of Living on West End," *New York Times*, February 23, 2003.
22 Frederick Douglass, *My Bondage and My Freedom*, qtd. in Rodgers, *Canaan*, 30.
23 Langston Hughes, *The Big Sea*, 81, qtd. in Rodgers, *Canaan*, 76.
24 *Canaan*, 72.
25 *Canaan*, 71.
26 Alain Locke, *The New Negro*, 6, qtd. in Rodgers, *Canaan*, 72.
27 Langston Hughes and Carl Van Vechten, *Remember Me to Harlem: The Letters of Langston Hughes and Carl Van Vechten*, ed. Emily Bernard (Vintage, 2002), 135; Gertrude Stein and Carl Van Vechten, *The Letters of Gertrude Stein and Carl Van Vechten, 1913–1946*, ed. Edward Burns (Columbia University Press, 1986), 471; Langston Hughes to Noel Sullivan in Cleveland (July 29, 1936), letter, Bancroft Library, MSS C-3 80, box 40, courtesy of Owen Walsh.
28 Alain Locke, "God Save Reality: Retrospective Review of the Literature of the Negro: 1936," *Opportunity: Journal of Negro Life* 14–15 (1936–37), 42.
29 Era Bell Thompson, *American Daughter* (University of Chicago Press, 1946), 288.
30 Juanita Harrison, letter to Mr. Clay, May 17, 1937, private collection.
31 Robert Stepto, *From Behind the Veil: A Study of Afro-American Narrative* (University of Illinois Press, 1979).
32 David M. Katzman, *Seven Days a Week: Women and Domestic Service in Industrializing America* (University of Illinois Press, 1981), 188.
33 Katherine Van Wormer, David W. Jackson III, and Charletta Sudduth, *The Maid Narratives: Black Domestics and White Families in the Jim Crow South* (Louisiana State University Press, 2012), 33.
34 Katzman, 3–43.
35 Katzman, 212.
36 Grossman, 26.
37 Elizabeth Stordeur Pryor, *Colored Travelers: Mobility and the Fight for Citizenship Before the Civil War* (University of North Carolina Press, 2016), 20.
38 Juanita Harrison to John Mason Brown, letter, n.d., John Mason Brown Papers, Houghton Library, Harvard University, Cambridge, Massachusetts.

39 Linda K. Kerber, *No Constitutional Right to Be Ladies: Women and the Obligations of Citizenship* (Farrar, Straus and Giroux, 1999), 68.
40 Ellery Sedgwick, *The Happy Profession* (Little, Brown, 1946), 211.
41 Helen Rose, Contributors' Column, *Atlantic Monthly* 156, no. 5 (November 1935), 48.
42 George Dickinson, Contributors' Column, *Atlantic Monthly* 156, no. 4 (October 1935), 38–39.
43 Dickinson.
44 Pryor, 6–7.
45 W. E. B. Du Bois, *Darkwater: Voices from Within the Veil* (Harcourt, Brace, 1920).
46 W. E. B. Du Bois, "Opinion," *The Crisis* 21, no. 5 (March 1921), 197; qtd. in Totten, 3.
47 Pryor, 19, 20.
48 W. E. B. Du Bois, *Dusk of Dawn: An Essay Toward an Autobiography of a Race Concept* (Harcourt, Brace, 1940), 58.
49 Miriam Thaggert, *Riding Jane Crow: African American Women on the American Railroad* (University of Illinois Press, 2022), 15.
50 Willi Coleman, "African American Women and Community Development in California, 1848–1900," 108, in *Seeking El Dorado: African Americans in California*, ed. Lawrence B. de Graaf, Kevin Mulroy, and Quintard Taylor (University of Washington Press, 2001), 98–126.
51 Lynn Hudson, *West of Jim Crow: The Fight Against California's Color Line* (University of Illinois Press, 2020), 19.
52 R. David McCall, "'Every Thing in Its Place': Gender and Space on America's Railroads, 1830–1899," MA thesis, Department of History, Virginia Tech, 1999, 41; Eve L. Ewing, new foreword, *Crusade for Justice, The Autobiography of Ida B. Wells*, by Ida B. Wells, 2nd ed. (University of Chicago, 2020), ix.
53 W. E. B. Du Bois, "Idlewild," *The Crisis* Vacation Number (August 1917), 169.
54 YWCA, "Mission Statement," https://www.ywca.org/about.
55 Katzman, 218–19.
56 Eileen V. Wallis, *Earning Power: Women and Work in Los Angeles 1880–1930* (University of Nevada Press, 2010), 62.
57 Thompson, *American*, 193.
58 Marcia Tremmel Goldstein, "Breaking Down Barriers: Black and White Women's Visions of Integration: The Young Women's Christian Association in Denver and the Phyllis Wheatley Branch 1915–1964," MA thesis, Department of History, University of Colorado Denver, 1995, 3.
59 Kimberly Crandall Bowling and Kriste Lindenmeyer, "How Did a Multi-Racial Movement Develop in the Baltimore YWCA, 1883–1926?" (State University of New York at Binghamton, 2003), https://documents.alexanderstreet.com/d/1000684612.
60 Thompson, *American*, 193.
61 Marcia Tremmel Goldstein, "Breaking Down Barriers: The Denver YWCA and the Phyllis Wheatley Branch, 1940 to 1949," 39, *Historical Studies Journal* [University of Colorado at Denver] 12 (1995), 35–69.
62 Adrienne Lash Jones, "Young Women's Christian Association," in *Black Women in America: An Historical Encyclopedia*, ed. Darlene Clark Hine et al. (Carlson, 1993), 1300.
63 Glenda Elizabeth Gilmore, *Gender and Jim Crow: Women and the Politics of White Supremacy in North Carolina, 1896–1920* (University of North Carolina Press, 1996), 193.

64 Thompson, *American*, 194.
65 "List of United States Citizens," S.S. *Calawaii*, Sailing from Los Angeles Harbor, January 14, 1925, Arriving at Port of Honolulu, January 24, 1925, "Hawaii, Honolulu Passenger Lists, 1900–1953."
66 Lee K. Davison and Carlos Ramirez, "Local Banking Panics of the 1920s: Identification and Determinants," 164, *Journal of Monetary Economics* 66 (2014), 164–77.
67 Goldstein, MA thesis, 2.
68 Goldstein, MA thesis, 12.
69 Goldstein, MA thesis, 66.
70 Goldstein, MA thesis, 67.
71 Goldstein, MA thesis, 68, 67.
72 Goldstein, MA thesis, 69.
73 Goldstein, MA thesis, 71.
74 Melanie Shellenbarger, "Lincoln Hills," *High Country Summers: The Early Second Homes of Colorado, 1880–1940* (University of Arizona Press, 2012), 126–50.
75 Claire Martin, "A Resort to Remember," *The Denver Post*, February 14, 2009.
76 Shereen Marisol Meraji and Laura Krantz, "During Segregation, a Mountain Oasis Gave Black Families a Summer Escape," *NPR*, August 16, 2015.
77 Goldstein, MA thesis, 60.
78 Nancy D. McCormick, *Saltair* (University of Utah Press, 1985).
79 Wallace Stegner, "Xanadu by the Salt Flats: Memories of a Pleasure Dome," *American Heritage* 32, no. 4 (June/July 1981).
80 "Democracy Defined at Moscow," *The Crisis* 54, no. 4 (April 1947), 105.

Chapter 3

1 P. Gabrielle Foreman, "Sankofa Imperatives: Black Women, Digital Methods, and the Archival Turn," 424, *a/b: Auto/Biography Studies* 38, no. 2 (2023), 423–35.
2 "Becomes Partner in Realty Firm: Warren McGrath," *Los Angeles Herald* 39, no. 238, July 5, 1913, p. 17.
3 Laura Redford, "The Promise and Principles of Real Estate Development in an American Metropolis: Los Angeles 1903–1923," PhD dissertation, Department of History, University of California, Los Angeles, 2014, 10.
4 Redford, 76.
5 "Subdivisions and Subdividers," *Los Angeles Times*, May 11, 1924, p. D3; "Dickinson Rites Set: Octogenarian Realty Man Will Be Paid Final Honor Tomorrow," *Los Angeles Times*, May 11, 1936, p. 23.
6 George Dickinson, Contributors' Column, *Atlantic Monthly* 156, no. 4 (October 1935), 38–39.
7 Lawrence Culver, *Frontier of Leisure: Southern California and the Shaping of Modern America* (Oxford University Press, 2010), 11.
8 Redford, 75.
9 Culver, 56.
10 Owen Walsh, "'Betwixt and Between': Juanita Harrison's Black Internationalist Practice," *Palimpsest: A Journal on Women, Gender, and the Black International* 10, no. 1 (2021), 5.

11 Lynn Hudson, *West of Jim Crow: The Fight Against California's Color Line* (University of Illinois Press, 2020), 2.
12 *The Crisis*, July 1913, qtd. in Josh Sides, *L.A. City Limits: African American Los Angeles from the Great Depression to the Present* (University of California Press, 2006), epigraph.
13 Sides, 12.
14 Scott Kurashige, *The Shifting Grounds of Race: Black and Japanese Americans in the Making of Multiethnic Los Angeles* (Princeton University Press, 2008), 34.
15 Gene Slater, "Op-Ed: How Los Angeles Pioneered the Residential Segregation That Helped Divide America," *Los Angeles Times*, September 10, 2021, https://www.latimes.com/opinion/story/2021-09-10/racial-covenants-los-angeles-pioneered.
16 Kurashige, 32, 34–35.
17 Redford, 103.
18 Redford, 103.
19 Kurashige, 27.
20 Kurashige, 18.
21 Douglas Flamming, *Bound for Freedom: Black Los Angeles in Jim Crow America* (University of California Press, 2006), 49.
22 Hudson, 15.
23 Mark Wild, *Street Meetings: Multiethnic Neighborhoods in Early Twentieth-Century Los Angeles* (University of California Press, 2008), 2.
24 Kurashige, 7.
25 Alison Rose Jefferson, *Living the California Dream: African American Leisure Sites During the Jim Crow Era* (University of Nebraska Press, 2020), 20.
26 Hadley Meares, "Sunkist Skies of Glory," *Los Angeles Curbed*, May 24, 2018.
27 Culver, 5, 3, 6.
28 Eileen V. Wallis, *Earning Power: Women and Work in Los Angeles 1880–1930* (University of Nevada Press, 2010), 68.
29 Gayle Gullett, "Narratives of Modernity and Black Women's Sexuality: African American Newspaper Women, Los Angeles, 1914," in City of Los Angeles (author), "Women's Rights in Los Angeles, 1850–1980," *Survey LA, Los Angeles Citywide Historic Context Statement*, October 2018, 66.
30 Flamming, 139.
31 "In Memoriam: Martha Daman Kellam," *The Record* (National City), May 2, 1895, p. 3.
32 "In Memoriam"; *The Daily Commonwealth*, September 28, 1882, p. 1.
33 US Census, 1870, Topeka, Kansas, FamilySearch, Entry for C C Kellam and Martha Kellam, https://www.familysearch.org/ark:/61903/1:1:MCJB-5J6.
34 *The Daily Commonwealth*, May 4, 1879, p. 3.
35 Steven Schoenherr, *Chula Vista Centennial: A Century of People and Progress, 1911–2011* (City of Chula Vista, 2011).
36 Obituary, Martha Daman Kellam, *The Record* (National City), April 25, 1895, p. 3.
37 "Letters to the Times: The Other Dickinson," *Los Angeles Times*, August 4, 1901, p. B6.
38 Kristin L. Hoganson, *The Heartland: An American History* (Penguin, 2019), 68.
39 D. E. Salmon, US Department of Agriculture, "Mexico as a Market for Purebred Beef Cattle from the United States" (Government Printing Office, 1902).
40 Display Ad 240, *Los Angeles Times*, February 29, 1920), p. V8.

41 US Census, 1920, Los Angeles, California, FamilySearch, entry for George W. Dickinson and Myra K. Dickinson, https://www.familysearch.org/ark:/61903/1:1:MHQL-XJR; US Census, 1930, Los Angeles, California, FamilySearch, entry for George W. Dickinson and Myra K. Dickinson, https://www.familysearch.org/ark:/61903/1:1:XCVP-6CZ.
42 *The Lance*, September 9, 1893, p. 4.
43 "Becomes Partner."
44 "Farewell Party," *Los Angeles Times*, May 18, 1924, p. 36.
45 "Large Party of Angelenos Sails on World Tour," *Los Angeles Times*, December 14, 1925, p. 10.
46 "Realty Head Is to Tour World," *Los Angeles Times*, December 6, 1925, p. E7; "Angelenos Will Sail Soon on World Jaunt," *Los Angeles Times*, October 5, 1925, p. 18; *Los Angeles Times*, November 11, 1925, p. A7.
47 F. Scott Fitzgerald, *The Great Gatsby* (Penguin, 1994), 132.
48 Kim Cooper, "The Baltimore Hotel, Empty No More," *Esotouric: Tours into the Secret Heart of Los Angeles*, August 24, 2018.
49 Cooper.
50 Charles Perry, "The Cafeteria: An L.A. Original," *Los Angeles Times*, November 5, 2003, p. 85.
51 Alexandra Rasic, "From the Homestead Kitchen: The Cafeteria Craze of Los Angeles, Part 2," Homestead Museum, February 17, 2021, https://homesteadmuseum.blog.
52 Grant Deed, September 8, 1925, Los Angeles County Registrar-Recorder/County Clerk. The deed states, "R.S. Shepherd & Ruth E. Shepherd, his wife, in consideration of Ten and no/100 Dollars, to them in hand paid, receipt of which is hereby acknowledged, do hereby Grant to Juanita Harrison, a single woman the real property in the City of Santa Monica, County of Los Angeles, state of California, described as Lot Three (3) in Block Two (2) of Erkenbrecher Syndicate Santa Monica Tract."
53 Culver, 6.
54 See Matthew Avery Sutton, *Aimee Semple McPherson and the Resurrection of Christian America* (Harvard University Press, 2009).
55 Jefferson, 21, 16.
56 Hudson, 208.
57 Hudson, chap. 6, "The Only Difference Between Pasadena and Mississippi Is the Way They're Spelled: Swimming in Southern California."
58 Jefferson, 27.
59 Marne L. Campbell, *Making Black L.A.: Class, Gender, and Community, 1850–1917* (University of North Carolina Press, 2016), 91.
60 US Census, 1930, Pasadena, California. FamilySearch https://www.familysearch.org/ark:/61903/1:1:XC8L-MBJ. Entry for David Cunningham and Alice B. Cunningham; US Census, 1940, Los Angeles, California. FamilySearch https://www.familysearch.org/ark:/61903/1:1:K9C1-K2D. Entry for David O. Cunningham and Alice F. Cunningham. California, County Marriages, 1849–1957. FamilySearch https://www.familysearch.org/ark:/61903/1:1:K88M-TKX. Entry for David O. Cunningham and Alice F. Grimes, 22 January 1922.
61 Jefferson, fig. 27, p, 104.
62 Juanita Harrison to Alice Foster, January 9, 1931, Ann Cunningham Smith collection. Hereafter parenthetically cited as FL1.

63 Obituary, Alice Foster Cunningham, "Last Rites Said for L.A. Pioneer," *Los Angeles Sentinel*, June 25, 1970, p. C11; Marriage License, Grimes and Cunningham.
64 Obituary, Cunningham.
65 "Girl Friends Club Sponsors Dance For 'Y' Benefit," *Los Angeles Sentinel*, June 30, 1949, p. A4.
66 Craig Robertson, *The Passport in America: The History of a Document* (Oxford University Press, 2012), 225.
67 Miriam Matthews, "Latest Book Reviews," n.d., clipping, Ann Cunningham Smith collection.
68 Robertson, 215–16, 223.
69 The library's "Shades of L.A." project ran from 1991 to 1997. "Photo Days" were held around the city, encouraging residents to bring in family photos to be copied and archived. https://exhibits.lapl.org/shadesofla/.
70 H. Vincent Price, "Dignitaries 'Ref' Bout," *Los Angeles Sentinel*, September 30, 1976, p. B2. Judge David F. Cunningham is not to be confused with Los Angeles City Councilman David Surmier Cunningham Jr. or his son, Los Angeles County Superior Court Judge David Cunningham III.
71 Harrison mentioned that she dedicated the book to both Dickinsons (FL3), but the volume only names Myra. Why this should be is one of those many small mysteries. Perhaps George didn't want his name to appear; perhaps her publishers wanted to underscore the text's identity as a women's project; perhaps Harrison did.
72 Katherine McKittrick, *Demonic Grounds: Black Women and the Cartographies of Struggle* (University of Minnesota Press, 2006), 54.

Chapter 4

1 Shirley Leckie, "Biography Matters: Why Historians Need Well-Crafted Biographies More Than Ever," 19, in *Writing Biography: Historians and Their Craft* (University of Nebraska Press, 2004), 4–26.
2 Marjorie Merrill Bliss, "Vibrant Juanita Doesn't Need Spelling or Grammar," *Des Moines Register*, May 31, 1936. p. 8.
3 Chimene Jackson, "Black Women's Journals Reflect Mine, Yours, and Ours: Through the Travel Writing of Juanita Harrison," in *Diary as Literature: Through the Lens of Multiculturalism in America*, ed. Angela R. Hooks (Vernon Press, 2020), 84.
4 A. W. M., "Odyssey of a Colored Woman," *The Province* (September 12, 1936), 48.
5 Judith Rollins, *Between Women: Domestics and Their Employers* (Temple University Press, 1985), 104–5.
6 Chimene Jackson, 82.
7 Risa Goluboff, *Vagrant Nation: Police Power, Constitutional Change, and the Making of the 1960s* (Oxford University Press, 2016), 1.
8 Goluboff, 116.
9 Goluboff, 117.
10 "More Slavery at the South, By a Negro Nurse," *The Independent* (January 25, 1912), 196–200, https://docsouth.unc.edu/fpn/negnurse/negnurse.html. Qtd. in Rhonda Bellamy and Bertha Boykin Todd, *Meet the Help: True Stories of Domestics* (Dorrance Publishing Company, 2015), 50.

11 Judith Madera, "Shaking the Basemap," in *The Black Geographic: Praxis, Resistance, Futurity*, ed. Camilla Hawthorne and Jovan Scott Lewis (Duke University Press, 2023), 52.
12 Smithsonian National Museum of American History, "Ocean Crossings: 1870–1969," in *On the Water*, https://americanhistory.si.edu/onthewater/exhibition/5_3.html.
13 Gary Y. Okihiro, *American History Unbound: Asians and Pacific Islanders* (University of California Press, 2015), 25.
14 Sowande M. Mustakeem, *Slavery at Sea: Terror, Sex, and Sickness in the Middle Passage* (University of Illinois Press, 2016), 7.
15 Mustakeem, 8.
16 William Pickens, "Wm. Pickens, Nearing Home, Writes Impressions of Trip 'Below Gulf of Mexico,'" *California Eagle*, September 25, 1936, p. 2.
17 Patricia Yaeger, *Dirt and Desire: Reconstructing Southern Women's Writing* (University of Chicago Press, 2000), 59.
18 Owen Walsh, "'Betwixt and Between': Juanita Harrison's Black Internationalist Practice," *Palimpsest: A Journal on Women, Gender, and the Black International* 10, no. 1 (2021), 14.
19 Walsh, "Betwixt," 10.
20 Patricia Yaeger, chap. 5, "Beyond the Hummingbird: Southern Gargantuas," in *Dirt and Desire*, 113–49.
21 Emily Lutenski, *West of Harlem: African American Writers and the Borderlands* (University Press of Kansas, 2015), 81.
22 Jean Toomer, *Cane* (Boni & Liveright, 1923), 145.
23 Korey Garibaldi, *Impermanent Blackness: The Making and Unmaking of Interracial Literary Culture in Modern America* (Princeton University Press, 2023), 81.
24 Mollie Godfrey and Vershawn Ashanti Young, "Introduction: The Neo-Passing Narrative," in *Neo-Passing: Performing Identity After Jim Crow*, ed. Mollie Godfrey and Vershawn Ashanti Young (University of Illinois Press, 2018), 17, 20.
25 Walsh, "Betwixt," 2.
26 Tim Youngs, "African American Travel Writing," in *The Cambridge Companion to Postcolonial Travel Writing*, ed. Robert Clark (Cambridge University Press, 2018), 117; Tim Youngs, "Pushing Against the Black/White Limits of Maps: African American Writings of Travel," *English Studies in Africa* 53, no. 2 (2010), 71.
27 Catherine Hobbs, "Introduction: Cultures and Practices of U.S. Women's Literacy," in *Nineteenth-Century Women Learn to Write*, ed. Catherine Hobbs (University of Virginia Press, 1995), 1–2.
28 Hobbs, 2.

Chapter 5

1 Booker T. Washington, "The Awakening of the Negro," *Atlantic Monthly*, September 1896, p. 78.
2 Emily Lutenski, *West of Harlem: African American Writers and the Borderlands* (University Press of Kansas, 2015), 115.
3 Brent Hayes Edwards, *The Practice of Diaspora: Literature, Translation, and the Rise of Black Internationalism* (Harvard University Press, 2003), 4.

4 Joan DeJean, *The Essence of Style: How the French Invented High Fashion, Fine Food, Chic Cafes, Style, Sophistication, and Glamour* (Simon and Schuster, 2007), 3.
5 Kevin Lane Dearinger, *Clyde Fitch and the American Theatre: An Olive in the Cocktail* (Rowman & Littlefield, 2016), 92.
6 US Census, 1920, New York City, FamilySearch, entry for Mary B Morris and Felice T Morris, https://www.familysearch.org/ark:/61903/1:1:MJYP-PZW.
7 "Pretty Good Remarks by the Press," *The Theater* 5 (1888), 423.
8 Petrine Archer, "Negrophilia, Diaspora, and Moments of Crisis," 33, in *Afro Modern: Journeys Through the Black Atlantic*, ed. Tanya Barson and Peter Gorschlüter (Harry N. Abrams, 2010), 26–39.
9 Archer, 33.
10 James Weldon Johnson, *Along This Way: The Autobiography of James Weldon Johnson* (Viking Press, 1933), 14.
11 T. Denean Sharpley-Whiting, "Langston Hughes and the Paris Transfer," *Arcade: The Humanities in the World*, https://shc.stanford.edu/arcade/colloquies/americans-paris.
12 Jessie Redmon Fauset, "Yarrow Revisited," *The Crisis*, January 1925, republished in *Comedy: American Style*, ed. Cherene Sherrard-Johnson (Rutgers University Press, 2009), 243; referenced in Totten, 91.
13 Jessie Redmon Fauset, Letter to Langston Hughes, qtd. in Totten, 96.
14 Jessie Redmon Fauset, "Biographical Note" in *Caroling Dusk: An Anthology of Verse by Negro Poets*, ed. Countee Cullen (Harper & Row Publishers, 1927).
15 Fauset, "Yarrow Revisited," 162.
16 Archer, 31.
17 T. Denean Sharpley-Whiting, *Bricktop's Paris: African American Women in Paris Between the Two World Wars* (State University of New York Press, 2015), 96.
18 John W. Graham, *Gold Star Mother Pilgrimages of the 1930s: Overseas Grave Visitations by Mothers and Widows of Fallen U.S. World War I Soldiers* (McFarland, 2005), 20.
19 Graham, 126.
20 Lisa M. Budreau, "Honoring Our Gold Star Mothers," from *We Return Fighting: World War I and the Shaping of Modern Black Identity* (Smithsonian Books, 2019), https://nmaahc.si.edu/honoring-our-gold-star-mothers.
21 Rebecca Jo Plant and Frances M. Clarke, "'The Crowning Insult': Federal Segregation and the Gold Star Mother Pilgrimages of the Early 1930s," *Journal of American History* 101, no. 2 (September 2015), 406–7.
22 Budreau.
23 Graham, 126.
24 Graham, 126.
25 Budreau.
26 Arthur Knight, *Disintegrating the Musical: Black Performance and American Musical Film* (Duke University Press, 2002), 78.
27 "Negro Mothers Forget Sorrows at Music Gala," n.d., clipping, Ann Cunningham Smith Collection.
28 George Dickinson, "Contributors' Column," *Atlantic Monthly* 156, no. 4 (October 1935), 38–39.
29 Halverson, "Epilogue: The *Atlantic* origins of *The Autobiography of Alice B. Toklas*," *Faraway Women and the "Atlantic Monthly"* (University of Massachusetts Press,

2019), 209–21. Stein added to the *Atlantic*'s pool of putatively plain memoirs through fashioning a deliberately plain voice. Even as *The Autobiography of Alice B. Toklas* was advanced by the "faraway women" genre, its recasting of the form helped sustain it

30 Mildred Morris, "Actors Say They Are Most Abused of All Workers," *The Evening Record*, Hackensack, August 15, 1919, p. 1.
31 "Irving R. Wiles: Distinctive American Portrait Painter," *The Craftsman* 18 (June 1910), 347.
32 May Day Lo, "Vagabond Writer, Artist Meet On Common Ground," *Honolulu Star-Bulletin*, May 29, 1937, p. 3.
33 William Norwood, *Honolulu Star-Bulletin*, March 25, 1936, p. 4.
34 "Author Indicates she, Like Garbo, Wants to Go Home," *Honolulu Advertiser*, August 13, 1937, p. 8.
35 Description, *Mildred Morris*, by Irving Ramsey Wiles, New York Historical Society Museum, https://emuseum.nyhistory.org/objects/26905/mildred-morris-18801966.
36 Carolyn Steedman, "Servants and Their Relationship to the Unconscious," 320, *Journal of British Studies* 42, no. 3 (2003), 316–50.
37 Harrison implied that the Dickinsons wrote her jointly, with a passing reference to letters "that He wrote for Mrs Dickinson" indicating that Myra may have dictated missives to George (FL3).
38 Miriam Matthews, "Latest Book Reviews," n.d., clipping, Ann Cunningham Smith Collection.
39 Juanita Harrison, Postcard to Mr. Clay, 1937, private collection.
40 Calvin Tomkins, "Living Well Is the Best Revenge," *New Yorker*, July 20, 1962.
41 F. Scott Fitzgerald, "How to Live on Practically Nothing a Year," *The Saturday Evening Post* 197, no. 2 (September 20, 1924), 165.
42 Frederick Douglass, *Life and Times of Frederick Douglass* (Park Publishing Co., 1881), 682, https://docsouth.unc.edu/neh/douglass/douglass.html.
43 Helen Rose, "Contributors' Column," *Atlantic Monthly* 156, no. 5 (November 1935), 48.
44 James Clifford, *Predicament of Culture: Twentieth-Century Ethnography, Literature, and Art* (Harvard University Press, 1988), 14, 4.
45 Susan L. Keller, "The Riviera's Golden Boy: Fitzgerald, Cosmopolitan Tanning, and Racial Commodities in *Tender Is the Night*," *The F. Scott Fitzgerald Review* 8 (2010), 130–59.
46 F. Scott Fitzgerald, *Tender Is the Night* (Charles Scribner's Sons, 1934), chap. 1.
47 Fitzgerald, "How to Live," 169.
48 Fitzgerald, "How to Live," 170.
49 Showing it had equal appeal to a Black middle class, Juan-les-Pins makes a cameo appearance in Jessie Redmon Fauset's 1933 novel, *Comedy: American Style*. The cruel Olivia Carey proposes the village as an ideal place for her light-skinned family to pass, so long as they jettison their dark son, Oliver. "I like Juan-les-Pins best," she imparts, "and they say property is marvelously cheap there" (160).
50 Sharpley-Whiting, *Bricktop's*, 8.
51 Fitzgerald, "How to Live," 165,
52 William Fortescue, *The Third Republic in France, 1870–1940: Conflicts and Continuities* (Routledge, 2000), 183.

53 Beth A. Simmons, *Who Adjusts: Domestic Sources of Foreign Economic Policy During the Interwar Years* (Princeton University Press, 1994), 257.

Chapter 6

1 Joyce E. Chapin, "Planetary Power? The United States and the History of Around-the-World Travel," *Journal of American Studies* 47, no. 1 (February 2013), 3.
2 Chapin, 12.
3 Miriam Matthews, "Latest Book Reviews," n.d., clipping, Ann Cunningham Smith Collection.
4 Valerie Popp, "Where Confusion Is: Transnationalism in the Fiction of Jessie Redmon Fauset," *African American Review* 43, no. 1 (2009), 136.
5 Camilla Hawthorne and Jovan Scott Lewis, "Introduction: Black Geographies: Material Praxis of Black Life and Study," in *The Black Geographic: Praxis, Resistance, Futurity*, ed. Camilla Hawthorne and Jovan Scott Lewis (Duke University Press, 2023), 5.
6 Alasdair Pettinger, "African Americans on Africa: Colleen J. McElroy and the Rhetoric of Kinship," *Journal of Transatlantic Studies* 7, no. 3 (2009), 317–28.
7 Brent Hayes Edwards, *The Practice of Diaspora: Literature, Translation, and the Rise of Black Internationalism* (Harvard University Press, 2003), 5.
8 Charles Forsdick and David Murphy, "France Must Acknowledge Its Colonial Past," *Guardian*, May 2, 2011; Nicolas Bancel, Pascal Blanchard, and Sandrine Lemaire, "Racist Theme Parks for Europe's Colonialists: Human Zoos," *Le Monde Diplomatique*, August 2000, https://mondediplo.com/2000/08/07humanzoo.
9 Forsdick and Murphy.
10 Birgitta Kuster, "Sous les yeux vigilants/Under the Watchful Eyes: On the International Exhibition in Paris 1931," translated by Aileen Derieg, *Transform*, May 2007, http://transform.eipcp.net/transversal/1007/kuster/en.html.
11 James Clifford, *Predicament of Culture: Twentieth-Century Ethnography, Literature, and Art* (Harvard University Press, 1988), 168–69.
12 Lynn Hudson, *West of Jim Crow: The Fight Against California's Color Line* (University of Illinois Press, 2020), 62.
13 Marina Magloire, "Speaking Through Signs: Juanita Harrison's Gendered Vagabondage," American Studies Association conference, Honolulu, November 2019.
14 Toqa Ezzidin, "Maison Groppi: Cairo's Downtown Historical Gem to Reopen Following Renovations," *Egyptian Streets*, November 2, 2017, https://egyptianstreets.com/2017/11/02.
15 Linda S. Heard, "If Only Groppi's Walls Could Talk," *Al Shindagah Magazine* 78 (October-November 2007), https://www.alshindagah.com/shindagah78/eng/groppi.htm.
16 W. E. B. Du Bois, "Of our Spiritual Strivings," in *The Souls of Black Folk: Essays and Sketches* (A.C. McClurg & Co., 1903).
17 "The Cotton Mills: Growth and Decline," *Mumbai Pages*, December 27, 1995, https://theory.tifr.res.in/bombay/history/cotton.html.
18 Jonathan Clarke, "'Like a Huge Birdcage Exhaled from the Earth': Watson's Esplanade Hotel, Mumbai (1867–71), and Its Place in Structural History," *Construction History* 18 (2002), 37–77.

19 "Rudyard Kipling to M.A. Jinnah: Mumbai's Crumbling Colonial homes," *Mint*, February 20, 2016, https://www.livemint.com.
20 T. V. Mahalingam, "How Mumbai's Watson's Hotel faded into Shades of Oblivion," *The Economic Times*, July 7, 2012, https://economictimes.indiatimes.com.
21 Mark Twain, *Following the Equator: A Journey Around the World* (American Publishing Co., 1897), 348.
22 Twain, *Following the Equator*, 352.
23 Twain, *Following the Equator*, 351.
24 Rudyard Kipling, *Plain Tales from the Hills* (Thacker, Spink and Company, 1888), 40.
25 The Taj Mahal Hotel's identity as a "symbol of Indian wealth and progress" made it the target of a deadly 2008 terrorist attack. Peter Foster, "Bombay Terror Attacks: Why the Taj Mahal Hotel Was Chosen," *Telegraph*, November 2008.
26 Clarke, 37.
27 Ian Strathcarron, *The Indian Equator: Mark Twain's India Revisited* (Courier Corporation, 2013), 8.
28 Nathaniel Knop, Peter Rippl, Ragunath Vasudevan, dirs., *The Watson's Hotel* (RV Films, 2019).
29 Chimene Jackson, "Black Women's Journals Reflect Mine, Yours, and Ours: Through the Travel Writing of Juanita Harrison," in *Diary as Literature: Through the Lens of Multiculturalism in America*, ed. Angela R. Hooks (Vernon Press, 2020), 77.
30 Victoria Lamont, "Li(v)es of a Woman Homesteader: Silence, Disclosure, and Self in the Letters of Elinore Pruitt Stewart," 224, *a/b: Auto/Biography Studies* 16, no. 2 (2001), 219–36.
31 Owen Walsh and Kate Dossett, "Gwendolyn Bennett and Juanita Harrison: Writing the Black Radical Tradition," *Comparative American Studies: An International Journal*, Volume 19, 1 (2022), https://doi.org/10.1080/14775700.2022.2026207; Tim Youngs, "African American Travel Writing," in *The Cambridge Companion to Postcolonial Travel Writing*, ed. Robert Clark (Cambridge University Press, 2018), 117.
32 Hilton Als, "The Elusive Langston Hughes," *New Yorker*, February 16, 2015.
33 Alex D. R. Castro, "Manila Carnivals: 1908–1939," July 16, 2008, http://manilacarnivals.blogspot.com/2008/07/2-man-plan-carnival.html.

Chapter 7

1 *Los Angeles Herald* 37, no. 266, June 24, 1910.
2 Jonathan Kinghorn, *The Atlantic Transport Line, 1881–1931: A History with Details on All Ships* (McFarland & Co., 2012), 229.
3 Hawaii, Honolulu, Passenger Lists, 1900–1954, FamilySearch, entry for Juanita Harrison, 1925, https://www.familysearch.org/ark:/61903/1:1:QVR9-KQTQ.
4 Sumner La Croix, "Economic History of Hawaii," *EH.Net Encyclopedia*, ed. Robert Whaples, September 27, 2001, https://eh.net/encyclopedia/economic-history-of-hawaii/.
5 John E. Reinecke, *The Filipino Piecemeal Sugar Strike of 1924–1925* (University of Hawaii Press, 1996), 1.
6 David E. Stannard, *Honor Killing: Race, Rape, and Clarence Darrow's Spectacular Last Case* (Penguin Random House, 2006), 2.
7 Nitasha Tamar Sharma, "The Racial Imperative: Reading Hawaii History and Black-Hawaiian Relations through the Perspective of Black Residents," in *Beyond Ethnicity:*

New Politics of Race in Hawaiʻi, ed. Camilla Fojas, Rudy P. Guevarra, and Nitasha Tamar Sharma (University of Hawaii Press, 2018), 125.
8. Stannard, 30, 2.
9. Tom Coffman, Inclusion: How Hawaiʻi Protected Japanese Americans from Mass Internment, Transformed Itself, and Changed America (University of Hawaii Press, 2021), 310.
10. Nancy Morris, "Hawaii Territory," American Heritage 56, no. 2 (April/May 2005), https://www.americanheritage.com/hawaii-territory.
11. Margaret Carrigan, "New Georgia O'Keeffe Show Is Full of Hawaii, Fruit and Questions," Observer, May 18, 2018, https://observer.com/2018/05/georgia-okeeffe-dole-ad-paintings-made-in-hawaii-have-been-reunited/.
12. See Alika Michael Bourgette, "Let's Talk Story: Waikiki and Its Social Displacements in Oral Histories and Print, 1901–1935," MA thesis, Department of History, California Polytechnic State University, 2017.
13. Bourgette, 117.
14. W. Somerset Maugham, "Honolulu" in Collected Short Stories (Penguin, 1951), 76.
15. "Volcano House" [advertisement], San Francisco Call, August 14, 1912, p. 14, https://chroniclingamerica.loc.gov/lccn/sn85066387/1912-08-14/ed-1/seq-14/.
16. Alan Bernheimer, "Mark Twain's 'Letters from Hawaii'" (1866), Nowhere Magazine, https://nowheremag.com/2020/12/mark-twains-letters-from-hawaii/.
17. "Guests register at Volcano House," Honolulu Star-Bulletin, June 3, 1925, p. 23.
18. "Legion Auxiliary Reception Honors Ladies of Fleet," Honolulu Star-Bulletin, May 16, 1925, p. 33.
19. "Australian Register of Historical Vessels," Australian National Maritime Museum, http://arhv.anmm.gov.au/events/180.
20. Mollie Godfrey and Vershawn Ashanti Young, "Introduction: The Neo-Passing Narrative," in Neo-Passing: Performing Identity After Jim Crow, ed. Mollie Godfrey and Vershawn Ashanti Young (University of Illinois Press, 2018), 21.
21. "List of United States Citizens," President Cleveland, Passenger Manifest, July 23, 1925, California, San Francisco, Passenger Lists, 1893–1953, FamilySearch, entry for Juanita Harrison, 1925, https://www.familysearch.org/ark:/61903/1:1:KX4R-6JY.
22. Miriam Matthews, "Latest Book Reviews," n.d., clipping, Ann Cunningham Smith Collection.
23. Bourgette, 2.
24. Linda K. Menton and Eileen Tamura, A History of Hawaiʻi, student edition (University of Hawaii Press, 1989), 169.
25. Eleanor C. Nordyke, "Blacks in Hawaiʻi: A Demographic and Historical Perspective," The Hawaiian Journal of History vol. 22 (1988), 242.
26. Barack Obama, Dreams from My Father (Times Books, 1995), 23.
27. Nitasha Tamar Sharma, Hawaiʻi Is My Haven: Race and Indigeneity in the Black Pacific (Duke University Press, 2021), 265.
28. Jonathan Y. Okamura, Raced to Death in 1920s Hawaiʻi: Injustice and Revenge in the Fukunaga Case (University of Illinois Press, 2019), 3.
29. Sharma, Hawaiʻi, 262.
30. Will Sabin, "Popular Misconceptions About Hawaii," Hawaiian Annual (1937), 52.
31. Edward D. Beechert, Working in Hawaii: A Labor History (University of Hawaii Press, 1985), 251.

32. La Croix.
33. "Prendergast Diamond Head Terrace Duplex Nomination [3020/3022 Hibiscus Drive]," Historic Hawaii Foundation, January 6, 2021, https://historichawaii.org.
34. James T. Hamada, "Juanita Harrison, *Atlantic Monthly* Writer Now Here, Is Mighty Fond of Japanese Baby," *Nippu Jiji*, April 22, 1936, p. 1.
35. Juanita Harrison, Contributors' Column, *Atlantic Monthly* (March 1937), 374.
36. Adria L. Imada, *Aloha America: Hula Circuits Through the U.S. Empire* (Duke University Press, 2012), 159.
37. Imada, 164.
38. Peter T. Young, "Lalani Village," Images of Old Hawaii, December 22, 2014, https://imagesofoldhawaii.com/lalani-village/.
39. Imada, 159.
40. "Waikiki Architecture: Royal Hawaiian Hotel," *The Hawaii Independent* (April 12, 2013).
41. Robert Trumbull, "Royal Hawaiian: Waikiki as It Was," *New York Times*, February 20, 1983, p. 60.
42. F. Scott Fitzgerald, *The Great Gatsby* (Penguin, 1994), 138–39.
43. *Honolulu Star-Bulletin*, October 30, 1937, p. 48.
44. Kurashige, 19.
45. "List of Manifest of Alien Passengers," August 30, 1918.
46. "List of Manifest of Alien Passengers," May 9, 1921.
47. US Census, 1940, Honolulu, FamilySearch, entry for Masakiki Tada and Kuni Tada, https://www.familysearch.org/ark:/61903/1:1:VB9X-DGC.
48. Peter T. Young, "James Steiner," Images of Old Hawaii, July 24, 2016, https://imagesofoldhawaii.com/james-steiner/.
49. "Steiner's Waikiki Mansion," *Honolulu Star-Bulletin*, August 10, 1912, p. 8.
50. "Japanese Remembered in Steiner Bequest," *Nippu Jiji*, June 8, 1939, p. 15.

Chapter 8

1. Juanita Harrison, letter to Mr. Clay, May 17, 1937, private collection.
2. "Umberto Tragella, Former Matson Official, Dies," *Honolulu Star-Bulletin*, October 11, 1968, p. 10.
3. Alma Whitaker, "Sugar and Spice," *Los Angeles Times*, August 2, 1937, p. A8.
4. Harrison, Contributors' Column, *Atlantic Monthly* (March 1937), 374.
5. Advertisement, *The Indianapolis Star*, May 17, 1936, p. 33.
6. Korey Garibaldi, *Impermanent Blackness: The Making and Unmaking of Interracial Literary Culture in Modern America* (Princeton University Press, 2023), 76.
7. Garibaldi, 91.
8. "In Brief Review," *The English Journal* 25, no. 7 (September 1936), 613.
9. John Kevin Young, *Black Writers, White Publishers: Marketplace Politics in Twentieth-Century African American Literature* (University Press of Mississippi, 2006), 19.
10. Garibaldi, 75.
11. *Atlantic Monthly* 156, no. 5 (November 1935), 601.
12. Judith Madera, "Shaking the Basemap," in *The Black Geographic: Praxis, Resistance, Futurity*, ed. Camilla Hawthorne and Jovan Scott Lewis (Duke University Press, 2023), 62.

13 William Brighty Rands, "Great, Wide, Beautiful, Wonderful World," in *Lilliput Lyrics* (John Lane, 1899), 101–2; as quoted in Kate Douglas Wiggin, *Rebecca of Sunnybrook Farm* (William Morrow, 1994), 119.
14 Qtd. in Susan L. Keller, "The Riviera's Golden Boy: Fitzgerald, Cosmopolitan Tanning, and Racial Commodities in *Tender Is the Night*," *The F. Scott Fitzgerald Review* 8 (2010), 139.
15 *El Paso Herald-Post*, May 22, 1936, p. 5.
16 Qtd. in "Miss Harrison's Book on Travels Is Well Received," *Nippu Jiji*, June 5, 1936, p. 5.
17 Langston Hughes to Noel Sullivan (July 29, 1936), Letter, Bancroft Library, MSS C-3 80, box 40, courtesy Owen Walsh.
18 Langston Hughes, "The Negro Artist and the Racial Mountain," in *Within the Circle: An Anthology of African American Literary Criticism from the Harlem Renaissance to the Present*, ed. Angelyn Mitchell (Duke University Press, 1994), 57, 58, 59.
19 Alain Locke, "God Save Reality: Retrospective Review of the Literature of the Negro: 1936," *Opportunity: Journal of Negro Life* 14–15 (1936–37), 42.
20 Arthur B. Spingarn, "Books by Negro Authors in 1936," *The Crisis* 44, no. 2 (1937), 48.
21 *Tampa Bay Times*, April 22, 1937, p. 9; *St. Cloud Times*, February 15, 1938, p. 6.
22 William Pickens, "Wm. Pickens, Nearing Home, Writes Impressions of Trip 'Below Gulf of Mexico,'" *California Eagle*, September 25, 1936, p. 2.
23 "Juanita Harrison, Author, Fetes Pickens in Waikiki Tent," *California Eagle*, August 28, 1936, p. 3.
24 May Day Lo, "Vagabond Writer, Artist Meet On Common Ground," *Honolulu Star-Bulletin*, May 29, 1937, p. 3.
25 "Miss Harrison's Book on Travels Is Well Received," *Nippu Jiji*, June 5, 1936, p. 5.
26 "Shirland Quin Gives Talk," *Honolulu Star-Advertiser*, November 1, 1936, p. 35.
27 "Juanita Harrison's Adventures," *Honolulu Advertiser*, December 31, 1935, p. 14; "City Scented by Garlands," *Honolulu Advertiser*, May 2, 1936, p. 7.
28 William Norwood, *Honolulu Star-Bulletin*, March 25, 1936, p. 4.
29 Martha Stermer, "The Diary of a Kona Farmer's Wife," *Honolulu Star-Bulletin*, July 7, 1937, p. 3.
30 Frances Lincoln, "The Skillful Journeyer," *Hawaii Tribune-Herald*, March 30, 1966, p. 17.
31 "City Scented by Garlands"; Advertisement, *Honolulu Star-Advertiser*, May 24, 1936.
32 Norwood.
33 James T. Hamada, "Juanita Harrison, *Atlantic Monthly* Writer Now Here, Is Mighty Fond of Japanese Baby," *Nippu Jiji*, April 22, 1936, p. 1; Gwen Dew, *My God: A Woman* (TransGate Enterprises, 2007).
34 Garibaldi, 80.
35 Hamada, 1.
36 Hamada, 1.
37 Ellery Sedgwick, *The Happy Profession* (Little, Brown, 1946), 215.
38 Sedgwick's portrait of Harrison little troubled his readership. None of the many positive reviews of *The Happy Profession* and none of the many congratulatory letters about it archived in his papers at the Massachusetts Historical Society object to the racial caricature or epithet.
39 "Juanita, Unique Authoress, Meets 'Unseen Sweetheart,'" *Honolulu Star-Bulletin*, March 25, 1936, p. 4.

40 Sedgwick, 215.
41 Adele Logan Alexander, introduction to *My Great, Wide, Beautiful World*, by Juanita Harrison (G.K. Hall & Co., 1996), xvii.
42 Sedgwick, 200.
43 "Juanita Harrison to Be Introduced," *Honolulu Advertiser*, August 12, 1937, p. 5.
44 "Author Indicates She, Like Garbo, Wants to Go Home," *Honolulu Advertiser*, August 13, 1937, p. 8.
45 "Mrs. Tufts in Review," *Honolulu Advertiser*, August 15, 1937, p. 33.
46 US Census, 1940, Los Angeles, FamilySearch, entry for Mary D. Tufts and Louise G. Leland, www.familysearch.org/ark:/61903/1:1:K9CZ-VLP.
47 "Ebell Club to Study Trends of Books and Events," *Long Beach Independent*, March 12, 1944, p. 22.
48 "Ebell Club to Study Trends," 22.
49 Whitaker.
50 "Mrs. Edward Tufts on Honolulu Vacation," *The San Bernardino County Sun*, August 4, 1937, p. 8.
51 "Juanita Harrison, Vagabond Author, to Make Local Bow," *Honolulu Star Advertiser*, August 10, 1937, p. 1; "Juanita Harrison to Be Introduced."
52 Harrison, postcard to Clay, Private collection.
53 US Census, 1940, Los Angeles, California, FamilySearch, entry for Myra K Dickinson, 1940, https://www.familysearch.org/ark:/61903/1:1:K9ZW-JNS.
54 "Memo Pad," *Los Angeles Times*, October 9, 1940, p. 29.
55 "Best Sellers of the Week, Here and Elsewhere," *New York Times*, August 24, 1936, p. 13; "Best Sellers of the Week, Here and Elsewhere," *New York Times*, June 29, 1936, p. 13.
56 Review of *My Great, Wide, Beautiful World*, *California Eagle*, June 26, 1936, p. 60.
57 "Miss Harrison's Book in Fifth Printing," *California Eagle*, August 14, 1936, p. 2.
58 "De Todo un Poco," *Opinión* 10, no. 317 (July 28, 1936).
59 Ruby Berkley Goodwin, "California's Negro Authors," *Los Angeles Sentinel*, June 25, 1959, p. B17.
60 "Anniversary of Library Is Celebrated," *California Eagle*, September 18, 1936.
61 Miriam Matthews, letter to Alice M. Foster, October 6, 1936, Ann Cunningham Smith Collection.
62 Miriam Matthews, "Latest Book Reviews," n.d., clipping, Ann Cunningham Smith Collection.
63 Matthews, letter.
64 Alexander MacDonald, *Honolulu Advertiser*, December 29, 1935, p. 29.
65 Leon Richards, "Multicultural-Multiracial Society or a Fragile Myth?," *Social Process in Hawaii* 43 (2014), 97.
66 Willis David Hoover, "Remember Pearl Harbor," *Honolulu*, December 7, 2016.
67 Jane L. Scheiber and Harry N. Scheiber, "Hawaii's Kibei Under Martial Law," *Western Legal History: The Journal of the Ninth Judicial Circuit Historical Society* 22 (2009), 18.
68 "War Bonds," *Nippu Jiji*, October 6, 1942, p. 4.
69 Obituary, George Y. Tada, *Augusta Chronicle*, July 31, 2016.
70 *Honolulu Star-Bulletin*, June 24, 1953, p. 13; *Honolulu Advertiser*, October 4, 1948, p. 5; *Honolulu Star-Bulletin*, November 26, 1953, p. 17.

71 *Honolulu Advertiser*, March 8, 1953, p. 31.
72 *Honolulu Star-Bulletin*, September 27, 1961, p. 28.
73 *Honolulu Advertiser*, November 1, 1942, p. 10.

Chapter 9

1 Louis A. Pérez, *On Becoming Cuban: Identity, Nationality, and Culture* (University of North Carolina Press, 1999), 166.
2 Adam Burns, "The Havana Special," February 23, 2022, https://www.american-rails.com/havana.html.
3 "FluentU," https://www.fluentu.com/blog/spanish/cuban-spanish/.
4 Pérez, 166.
5 Pérez, 167.
6 Pérez, 173, 172.
7 Pérez, 180, 179, 181.
8 Kevin Grogan, "Cuba's Dance of the Millions: Examining the Causes and Consequences of Violent Price Fluctuations in the Sugar Market between 1919 and 1920," MA thesis, Department of Latin American Studies, University of Florida, 2004, 17.
9 Pérez, 93.
10 Pérez, 193.
11 Qtd. in Emily Lutenski, *West of Harlem: African American Writers and the Borderlands* (University Press of Kansas, 2015), 117.
12 According to the 1920 museum catalog, *Las Meninas* is the only portrait of King Philip's family that the Prado Museum held at the time, and its size and fame make it the likely candidate to be displayed in a room of its own. That Harrison identifies two children in the painting might be due to her mistaking one of the attendants, a young man of short stature, for a child.
13 Michel Foucault, *The Order of Things: An Archeology of the Human Sciences* (Pantheon Books, 1970), 5.
14 Jefferey M. Sellers, review, *Exporting Japan: Politics of Emigration to Latin America* (University of Illinois Press, 2009), by Toake Endoh, *Tulsa Comparative Political Studies* 43, no. 2 (2010), 260–71; Kobe YWCA, https://www.kobe.ywca.or.jp/top/about.
15 Endoh, 18.
16 Sellers, 261.
17 Brasil, Cartões de Imigração, 1900–1980, FamilySearch, entry for Juanita Harrison, 1940, www.familysearch.org/ark:/61903/1:1:KCXW-WCT.
18 Michiel Baud and Annelou Ypeij, "Cultural Tourism in Latin America: An Introduction," in *Cultural Tourism in Latin America: The Politics of Space and Imagery*, ed. Michiel Baud and Annelou Ypeij (Brill, 2009), 7, 2.
19 Valerie Popp, "Where Confusion Is: Transnationalism in the Fiction of Jessie Redmon Fauset," 136, *African American Review* 43, no. 1 (2009), 136.
20 Popp, 132.
21 "Fiction in Lighter Vein," *New York Times*, September 27, 1942, p. BR32.
22 *Honolulu Advertiser*, November 1, 1942, p. 10.
23 "Time to Challenge Argentina's White European Self-Image," *Guardian*, May 31, 2021.

24 "Time."
25 Era Bell Thompson, *American Daughter* (University of Chicago Press, 1946), 288.
26 Thompson, "Argentina: Land of the Vanishing Blacks," *Ebony*, October 1973, p. 74.
27 Thompson, "Argentina," 74. The paths of these two world citizens, Harrison and Baker, frequently crossed.
28 James W. McGuire, *Peronism Without Perón: Unions, Parties, and Democracy in Argentina* (Stanford University Press, 1997).
29 Ellery Sedgwick, *The Happy Profession* (Little, Brown, 1946), 213.
30 After martial law was lifted in Honolulu, perhaps Harrison voiced some interest in returning, leading to Sedgwick's assumption that she actually did. Just as likely: He played hard and fast with the facts for the sake of a tidy story, as was his wont.

Chapter 10

1 Harrison, postcard to Clay, private collection.
2 Juanita Harrison, application for Passport, US consulate, Buenos Aires, February 17, 1950.
3 John Bengston, "Rare Chaplin Scenes in Downtown Los Angeles," June 14, 2011, https://silentlocations.com.
4 George Dickinson's grandson, also named George Dickinson, made the news in 2012 as one of two trustees accused of questionable dealings with their advisee, Susan Strong Davis. Once a famous ice skater and socialite, Davis was the daughter of the elder George Dickinson's long-time partner, Frank Strong. In her dotage she was found to have donated an enormous sum to a religious organization to which she had no known connection. Harriet Ryan, "Widow with Dementia Gave $600,000 to Kabbalah Centre Charity," *Los Angeles Times*, April 9, 2012, https://www.latimes.com/archives/la-xpm-2012-apr-09-la-et-kabbalah-larkin-2012 0409-story.html.
5 "Presidential Statement Signing the Social Security Act," August 14, 1935, *Social Security History*, https://www.ssa.gov/history/index.html.
6 Harrison, letter to Clay.
7 Master Property Records, County of San Diego, 7316 Encelia Drive, La Jolla, California 92037, 1952.
8 "California, Los Angeles Passenger Lists, 1907–1948," entry for Arthur T. Heuckendorff, 1932; "Hawaii, Honolulu Passenger Lists, 1900–1953," entry for Arthur T. Heuckendorff, 1926.
9 "Roger Guillemin—Biographical," NobelPrize.org, Nobel Prize Outreach AB 2023, https://www.nobelprize.org/prizes/medicine/1977/guillemin/biographical/.
10 Russell C. Ruffing, Historic American Engineering Record, Venice Canals HAER No. CA-124, City of Los Angeles, Bureau of Engineering, February 1992, www.venicenc.org/assets/documents/5/committee62f19969923a9.pdf.
11 Mark Twain, "A Prose Poem on Hawaii" (1889), Honolulu: Mercantile Press, ca. 1920, Beinecke Rare Book and Manuscript Library, https://collections.library.yale.edu/catalog/2004872?child_oid=1024449.
12 Barack Obama, *Dreams from My Father* (Times Books, 1995), 23.

13 Sumner La Croix, "Economic History of Hawaii," *EH.Net Encyclopedia*, ed. Robert Whaples, September 27, 2001, https://eh.net/encyclopedia/economic-history-of-hawaii.
14 Martin Luther King Jr., "Address to the House of Representatives of the First Legislature, State of Hawaii, on September 17, 1959," The Martin Luther King, Jr. Research and Education Institute, Stanford University, https://kinginstitute.stanford.edu/king-papers.
15 James T. Hamada, "Juanita Harrison, *Atlantic Monthly* Writer Now Here, Is Mighty Fond of Japanese Baby," *Nippu Jiji* (April 22, 1936), 1.
16 Alice Walker, "In Search of Zora Neale Hurston," *Ms. Magazine* (March 1975), 74–89.
17 "We'll Tell You," *Honolulu Advertiser*, July 26, 1942, p. 7.
18 William Norwood, *Honolulu Star-Bulletin* (March 25, 1936), 4.
19 James Weldon Johnson, *The Autobiography of an Ex-Colored Man* (Sherman, French, and Co., 1912).
20 Chimene Jackson, "Black Women's Journals Reflect Mine, Yours, and Ours: Through the Travel Writing of Juanita Harrison," in *Diary as Literature: Through the Lens of Multiculturalism in America*, ed. Angela R. Hooks (Vernon Press, 2020), 77.

Epilogue

1 John Mason Brown, "Seeing Things Topsy Turvy," *Saturday Review of Literature*, October 6, 1945, p. 25.
2 Brown, 24.
3 Juanita Harrison to John Mason Brown, letter, n.d., John Mason Brown Papers, Houghton Library, Harvard University, Cambridge, Massachusetts.
4 Arthur B. Spingarn, "Books by Negro Authors in 1936," *The Crisis* 44, no. 2 (1937).
5 Chimene Jackson, "Black Women's Journals Reflect Mine, Yours, and Ours: Through the Travel Writing of Juanita Harrison," in *Diary as Literature: Through the Lens of Multiculturalism in America*, ed. Angela R. Hooks (Vernon Press, 2020), 86.

Index

Page numbers in italic indicates an illustration.

Absalom, Absalom! (Faulkner), 189
Adams, Maude, 118, *126*
African American women: and activism, 53–54; domestic service, 33, 50, 51; geography, 7, 87; *My Great, Wide, Beautiful World*, 61, 120, 122, 206
African Americans: and activism, 50–51, 53–54, 77; archive, 10; leisure, 33, 54, 66, 77; literature, 9, 49, 92; mobility, 50–51; travel, 3, 54; in Chicago, 44, 45; Denver, 57; Honolulu, 172; Los Angeles, 66; New York, 46–47; Paris, 7, 119–20
Ala Moana trial, 166
Alaska, 207
Alexander, Adele Logan, 199
Als, Hilton, 258n32
Ambrosius, Lloyd E., 12
American West, 49, 58
Ann Cunningham Smith collection, 79–80, 81–82
Antin, Mary, 199
antisemitism, 5, 112
Archer, Petrine, 120
Argentina, 222–26
Armitage, Susan, 42
artists, 124, 131, 135–36
Atlantic Monthly and: Harrison 2, 5, 10, 51–52, 81, 115, 119, 124, 129, 173, 174–75, 177, 186–87, 188, 189, 192, 196, 203; George Dickinson, 52, 63, 123; Mildred Morris, 119; Helen Rose, 52, 135; Gertrude Stein, 124, 256n29
Autobiography of Alice B. Toklas, The (Stein), 124, 256n29

"B.," Mr. and Mrs., 216–17
Baker, Josephine, 81, 120, 224, 264n27
Baltimore Hotel, 74–75
bank: account, 231, 233; failure, 57, 62, 164
Barcelona International Exposition, 144, 216
Barcelona, 51, 96, 210, 214, 215, 216
Barson, Tanya, 255n8
Baud, Michiel, 264n17
Baughn, Jennifer, 246n38
Beechert, Edward D., 260n31
Bellamy, Rhonda, 247nn53–54
Bellamy, Rhonda, 254n10
Bengston, John 264n3
Bennett, Gwendolyn, 120
Bernard, Emily, 248n27
Bernheimer, Alan, 259n16
Birmingham (Alabama), 33, 35, 43
Blake, Tom, 245n8, 245nn10–11
Blanding, Don, 177, 203
Bledsoe, Adam, 244n20
Bliss, Marjorie Merril, 253n2
Boal, Mary, 70
Bolivia, 42, 226
Bontemps, Arna, 42
Bourgette, Alika, 171, 259nn12–13
Bowling, Kimberly Crandall, 249n59
Bradshaw's Continental Railway Guide, 112
Brandien, Carl, 127, 194–95
Brazil, 208, 218, 220, 221, 222
Brown, Charlotte, 54
Brown, John Mason, 16, 238, 239, 240
Brown, Lois, 10, 11
Budreau, Lisa M., 255n20

INDEX

Buenos Aires, 223–24, 224–26
burial, 234–35
Burma, 72, 113, 117, 152–53, 206
Burns, Adam, 263n2
Burns, Edward, 248n27
Burton, Jennifer, 5
Burton, Richard F., 154

"C.," Captain and Mrs., 121, 122
Cairo, 95, 98, 105, 124, 150–51, 175, 225
California Eagle, 26, 100, 194, 205
Camp Nizhoni, 59
Camp, Stephanie, 7, 27
Campbell, Marne L., 252n59
Canada, 211–12
Cane (Toomer), 109
Carillo, Ramón, 224
Carrigan, Margaret, 259n11
Castro, Alex D. R., 258n33
Catholicism, 213, 239, 241
Chamberlain, Ava, 11
Chapin, Joyce, 141
Chicago Defender, 44
Chicago World's Fair, 44, 73
childhood, 20, 21, 28–36
Chile, 222
China, 92–93, 123
Choctaw, 21, 22, 24–25, 40
circumnavigation, 74, 141–42, 164
Claire, Martin, 250n75
Clarke, Frances M., 122
Clarke, Jonathan, 258 n18
Clay, Mr., 185, 204, 211
Clifford, James, 137, 143
climate, 93, 116, 171, 228
clothing, 28, 33, 127, 151–52, 190, 198–99
Coffman, Tom, 259n9
Coleman, Willi, 249n50
Colored Young Women's Christian Association, 56
Columbus (Mississippi), 22, 27
Comedy: American Style (Fauset), 120, 256–57n49
Conroy, Jack, 42
Coontz, Admiral Robert, 169–70
Cooper, Kim, 252nn48–49
correspondence, 97–98, 112–14, 123, 128; with: employers, 2, 9, 39, 118, 130–31, 151; Dickinsons, 66, 79, 86–87, 256n37; Alice Foster, 12, 28, 30–31, 32, 61, 69, 74, 77, 79–82, 84, 85, 86, 87, 88, 106, 122, 124, 128–29, 130–31, 153, 181, 199, 203, 206, 207, 228; Mississippians, 32, 44, 59, 87; Mary Morris, 114, 118, 131; Helen Rose, 131
cosmopolitanism, 92, 142, 171, 177
countercartography, 8, 93
Crigler family, 21, 22–23
Crigler, Angela, 246n18
Crigler, Dr. John Lewis, 23
Crigler, Lizzie, 21, 23–24, 30
Crigler, Rosa, 21
Crisis, 53, 60, 119, 251n12
Cuba, 115, 116, 136, 170, 211–14, 215
Culley, Margo, xi
Culver, Lawrence, 64, 68, 251n9
Cunningham, Alice Foster, 78, 81–83
Cunningham, David O., 78
Cunningham, Judge David Foster, 84
Cutter, Martha, 244n32

Darjeeling, 152
Davis, Frank Marshall, 188
Davis, Susan Strong, 264n4
Davison, Lee K., 250n66
de Graaf, Lawrence B., 249n52
Dearinger, Kevin Lane, 255n5
death, 233, 237
DeJean, Joan, 117
Delany, Martin, 221–22
Denver, 49, 57–59, 69, 170
Detroit, 44–45
Dew, Gwen, 194, 195–96
Diamond Head Terrace, 176–77
Díaz, Porfirio, 71
Dickinson, Charles, 71
Dickinson, Colonel William Green, 70
Dickinson, George and Myra and: country club, 67; family, 69–71; Harrison, 61, 62, 63–64, 66, 69, 72–73, 84–88, 201, 228; real estate company, 62, 63, 70–71, 72; restrictive covenants, 62, 65; travels, 73–74; in Hawaii, 164; in Mexico, 71–72. *See also* correspondence
Dickinson, George (grandson), 264n4
Dickinson, Martha, 71

Dickinson, Myra, 88, 123, 187, 188, 203, 204–5, 229, 253n71
Dickinson, Wallace, 70
Dickinson, William, 71, 229
Dillingham, Walter F., 169, 176, 198
Djibouti, 143, 146–49
dog, 181, 202
domestic service, 35–37, 50–52, 96–99, 131, 132, 157, 186, 215, 218
domesticity, 101–2, 103
Douglass, Frederick, 28, 47, 133
Du Bois, W. E. B., 7, 53, 54, 64, 152, 156
Dyer, Richard, 190

Eaves, Latoya E., 244n20
education, 29–30, 55
Edwards, Brent Hayes, 117, 143
employers: relationships with, 37, 38, 51, 52, 97, 117–18, 121, 127, 135, 177, 202; in Honolulu, 176–77; in Mississippi, 29, 32; overseas, 36, 50, 96; on Riviera, 134–37; in Spain, 215–18
Endoh, Toake, 263n15
Eskridge, Robert, 203
Ewing, Eve L., 249n52
Ezzidin, Toqa, 257n14

family, 20–22, 30–31
"Faraway Women," 199, 221
Fauset, Jessie Redmon, 3, 46, 120–21
Fernández, Alberto, 224
Fitzgerald, F. Scott: "How to Live on Practically Nothing a Year," 133, 134, 137–38, 139; *The Great Gatsby*, 74, 181; *Tender is the Night*, 133, 138
Flamming, Douglas, 251n21
flappers, 137, 148
Florida, 48, 211
Foreman, P. Gabrielle, 61
Forsdick, Charles, 257nn8–9
Fortescue, William, 257n52
Foster, Alice and: family, 61, 78–79; Harrison, 61, 79–82, 130, 203, 206, 207. *See also* correspondence
Foster, Edna, 78, 82
Foster, Lawrence R., 78
Foster, Peter, 258n25
Foucault, Michel, 218

freedmen, 24, 25, 40
freedom, 3, 99, 103
French Riviera, 116, 131–40, 177, 179, 207, 218
Fuentes, Marisa, 20–21

Gardner, Eric, 10
Garibaldi, Korey, 5, 23, 109, 188, 189, 196, 243n4
Gates, Henry Louis, 5
George W. Dickinson & Co, 70–71, 72, 229
Gilmore, Glenda, 56–57
Godfrey, Mollie, 109, 244n32, 245n36
Gold Star Pilgrimage, 81, 121–22
Goldstein, Marcia Tremmel, 58, 249n61, 250nn69–74, 250n77
Goluboff, Risa, 98
Gone With the Wind (Mitchell), 189
Goodwin, Ruby Berkley, 205
Gorschlüter, Peter, 255n8
Graham, John W., 122
Graham, John W., 255nn18–19
Grand Tour, 8
grave, 234, 235
Great Migration, 3, 40, 41–42, 43, 66
Great Salt Lake, 59, 60
Grogan, Kevin, 212
Groppi's teahouse, 150–51
Gross, Kali Nicole, 6
Grossman, James, 44, 248nn14–15, 248n18, 248n36
Gruesz, Kirsten Silva, 214
Guillemin, Roger C., 230
Gullett, Gayle, 251n29

Hall, Michael Ra-Shon, 4
Halverson, Cathryn: *Faraway Women and the "Atlantic Monthly*," 199, 244n27, 256n29; *Maverick Autobiographies*, 244n27; *Playing House in the American West*, 101, 244n27
Hamada, James, 196–98, 246n30
Hamlet, Obrey Wendell, 58
Hardison, Ayesha K., 11, 32
Harlem Renaissance, 6, 46–47, 48, 117, 235–36
Harris, Odette, 97

Harrison, Jones, 21
Hartman, Saidiya, 11, 41
Havana, 211, 212–13, 214
Hawaii, 163–205, 207–9; demographics, 172, 182, 236; history, 165–66, 172; tourism, 166–68, 232–33
Hawthorne, Camilla, 142, 244n26, 245n3, 254n11, 257n5, 261n12
Heard, Linda S., 257n15
Herr, Maggie, 70
Heuckendorff, Arthur T. and Ruth R., 230
Higgons, Ruth, 73
Hobbs, Catherine, 112
Hoganson, Kristin L., 252n38
Honolulu, 13, 25, 49, 69, 81, 91, 102, 127, 228, 231–36. *See also* Hawaii
Hooks, Angela R., xiv, 243n11, 246n32, 253n3, 258n29, 265n20 (chap. 10), 265n5 (epilogue)
Hoover, Willis David, 208
Hudson, Lynn, 54, 66, 251n11
Hughes, Langston, 44, 46, 47, 48, 58, 192, 193
Hurst, Fannie, 189
Hurston, Zora Neale, 1, 10, 26, 58, 189, 234–35

identity: American, 106–7, 136–37, 147, 214, 219–20; authorial, 81, 157, 185, 186, 202, 225; Californian, 76, 107–8; class, 98–99; racial, 32, 37–38, 80, 109–10, 156, 157, 195–96, 236–37, 241; Southern, 37–38, 108–9, 110–11. *See also* Native heritage
Imada, Adria L., 180
Imitation of Life (Hurst), 189
Immigration Act of 1924, 168, 183
India, 152–57
investments, 63–64, 65–66, 75–76, 83, 85, 93, 106, 123, 235, 252n52
Iowa, 49–50
itinerary, 92–93, 95, 121

Jackson, Chimene, 4, 26, 94, 97, 156, 237, 241
Jackson, David H, Jr., 247n40
Jackson, David W., III, 248n43
Jackson, Helen Hunt, 210
Jackson, Leon, 13

Jackson, Nancy Beth, 248n21
James Cole, 245nn8–9
Jane Crow, 11, 53
Japan, 158–59, 160, 180–82
Japanese: and emigration, 183, 219; in Hawaii, 165, 175–76, 182–83, 208–9
Jardin d'Acclimatation, 1, 142, 143–44, 237
Jefferson, Alison Rose, 68, 77, 84, 251n25
Jerkins, Morgan, 44
Johnson, Charles S., 43
Johnson, James Weldon, 41, 119, 171, 236
Jolson, Al, 122
Jones, Adrienne Lash, 249n62
Juan-les-Pins, 55, 131–32, 137, 256n49

Kahahawai, Joe, 166
Kalākaua, King David, 154, 175, 181
Katzman, David, 32, 37, 50, 247n44, 247nn49–52, 248n32, 249n55
Kellam, Charles, 69–70
Kellam, Martha, 69–70
Keller, Susan, 138
Kennedy, President John F., 166
Kerber, Linda K., 33, 249n39
King, Martin Luther, Jr., 233
Kinghorn, Jonathan, 258n2
Kinney, Abbot, 230
Kipling, Rudyard, 154–55
Knight, Arthur, 256n26
Knop, Nathaniel, 258n28
Kobe, 55, 106, 159, 160, 182, 218–19
Korea, 158
Krantz, Laura, 250n76
Krauthamer, Barbara, 22, 24, 246n20, 246n22, 246n28
Kuluwaimaka, James Kapihe Palea, 180
Kurashige, Scott, 64–65, 251nn19–20, 251n24
Kuster, Birgitta, 257n10
Kyoto, 182

La Croix, Sumner, 258n4, 260n32, 265n13
La Jolla, 227, 228, 229–30
Lalani Hawaiian Village, 179–80
Lamont, Victoria, 157
language: English, 94; French, 115, 150, 170, 213, 214; Spanish, 211, 214, 220
Larsen, Nella, 47

Lau, Pat, 234
Leavell, R. H., 248n12
Leckie, Shirley, 91–92
leisure, 6, 9, 33, 58–59, 77, 98–99
Lepore, Jill, 10
Lewis, Jovan Scott, 142, 244n26, 245n3, 254n11, 257n5, 261n12
Lili'uokalani, Queen Lydia, 167, 175
Lincoln, Frances 261n30
Lindenmeyer, Kriste, 249n59
literacy, 29, 111–12, 215
Lo, May Day, 256n32, 261n24
Locke, Alain, 46, 48, 193
London, Jack, 169
López, Ian Haney, 11
Los Angeles, 49, 61–70, 72–79, 82–84, 88, 115, 164, 201, 228
Louisiana, 116–17, 214
Lowndes County, 22, 27, 41
Lutenski, Emily, 109, 116, 247n7, 263n11

MacDonald, Alexander, 263n64
Macmillan, 188, 205, 223
Madera, Judith, 7, 8, 10, 87, 99, 189, 246n33
Madrid, 13, 214, 216–18
Magloire, Marina, 144
Mahalingam, T. V., 258n20
Manila Carnival, 159
Mann, Thomas, 193
Mar del Plata, 209, 222, 223
Marks, Carole, 247n3, 247n5
Massie, Thalia, 166
Matson Navigation Company, 176–77, 180
Matthews, Miriam, 78–79, 82–83, 130, 205–7
Maugham, W. Somerset, 167–68
Mayo, Katherine, 107
McCall, R. David, 249n52
McCalph, Maude, 72
McCormick, Nancy D., 250n78
McCormick, Stacie, 245n37
McFarland, Katherine, 203
McGuire, James W., 264n28
McKee, Ruth, 203
McKittrick, Katherine, 7, 87
McPherson, Aimee Semple, 77
Meares, Hadley, 251n26
Meninas, Las, 13, 217–18

Menton, Linda K., 259n24
Meraji, Shereen Marisol, 250n76
Mississippi, 19–38, 41, 125
Mitchell, Dennis J., 246nn25–26
Mitchell, Margaret, 189
Moana Hotel, 177, 184, 185, 233
"More Slavery at the South," 36–37, 98
Morgan, Julia, 175
Morris, Felice, 118, 124, 127, 140
Morris, Felix, 118
Morris, Mary, 106, 113, 118, 120, 121, 122, 124, 130, 140, 187
Morris, Mildred and: career (acting), 13, 118, 125, 126; career (writing), 118–19, 125, 140; education, 125; preface to *My Great, Wide, Beautiful World*, 13, 28, 30, 33, 37, 46, 52, 55, 62, 69, 110, 115, 124–25, 127, 164, 189, 190, 192; work with Harrison, 16, 28, 30, 39, 79, 94, 114, 118, 119, 120–21, 123–25, 127–29, 130, 188, 242
Morris, Nancy, 259n10
Mossman, George, 179
Mossman, Pualani, 180
Mulroy, Kevin, 249n52
Mumbai, 13, 47, 87, 97, 101–2, 152, 153–56, 169
Murphy, David, 257nn8–9
Murphy, Gerald and Sara, 132–33
Mustakeem, Sowande, 100
My Great, Wide, Beautiful World (Harrison) and: composition, 9, 94, 123, 127–29, 129–31, 157; contract, 130, 174, 186; cover, 190, *191*; dedication, 86, 253n71; editing, 94, 128; epilogue, 103, 164, 173, 177, 179; gender, 103–6, 148; genre, 9, 40; Great Migration, 40; language, 94, 108; marketing, 111, 163, 188, 203, 205; paratexts, 189; preface, 28–29, 30, 37, 46, 62, 69, 110, 115, 118, 124–25, 127, 164, 189, 190, 192 (*see also* Mildred Morris); reception, 5, 9, 188, 194: by African Americans, 6, 192–93, 205, 206; in Honolulu, 195–96, 199, 205; in Los Angeles, 203, 205–7; reviews, 2, 5, 94, 95, 111, 192, 205, 206; sales, 2, 187, 188, 203, 205, 226; and the South, 108–9; structure, 2, 94, 97, 123, 189–90; title, 189–90; on work, 95–99

"My Great, Wide, Beautiful, Wonderful World" (Rands), 189–90

Nail family, 25
Nail, Manerva, 25
name, 26, 210, 245n38
National City, 70–71, 229
Native heritage, 24–26
Negro Motorist Green-Book, The, 54, 188
New York City, 43, 44, 45–48, 69, 91, 117, 118, 125, 216
Nice, 65–66, 109, 129, 132, 133–34, 171
Nippu Jiji, 192, 195, 196–98, 208
Nordyke, Eleanor C., 259n25
Northwood, Enid, 195
Norwood, William, 195, 243n3, 246n30, 256n33, 265n18
Noxolo, Pat, 4

O'Brien, Greg, 245n14
O'Keeffe, Georgia, 166
Obama, Barack, 172, 232
Okamura, Jonathan Y., 260n28
Okihiro, Gary Y., 99
Oktibbeha County, 21, 22, 23
Ordish, Rowland Mason, 154
Orientalism, 29, 151, 190

Pablo, Picasso, 133
package tours, 59
Paradise Lost, 45, 238
parents, 21
Paris Exposition, 143, 147
Paris, 37, 93, 116–22, 129, 131, 132, 134, 140, 145
Parish, Benjamin, 23
Parkridge Country Club, 79
Parks, Frederick, 246n18
Pasadena, 78
passing, 13, 109, 110, 118, 170, 224. *See also* identity
passport, 20, 42, 83–84, 106, 198, 206, 220, 222, 225, 226, 237
Pérez, Louis A., 212
Perón, Juan, 8, 225
Perry, Charles, 252n50
Peter Pan, 118, 126

Pettinger, Alasdair, 143
Philippines, 159–60
photographs, 151–52, 190
Phyllis Wheatley clubs, 56, 57–58
Pickens, William, 7, 100–101, 164, 194
Pieschel, Bridget Smith, 247n41
Plant, Rebecca Jo, 122
Pleasant, Mary, 54
Popp, Valerie, 142, 221–22
poste restante, 81, 93, 113, 114, 152
Price, H. Vincent, 253n70
Pryor, Elizabeth Stordeur, 3, 30, 50–51, 53

Quicksand (Larsen), 47

Ramirez, Carlos, 250n66
Rands, William Brighty, 190
Rasic, Alexandra, 252n51
Redford, Laura, 65, 250nn3–4, 251n8, 251n18
Reinecke, John E., 259n5
reputation, 31, 33, 104
Richards, Leon, 263n65
Rippi, Peter, 258n28
Ritterhouse, Jennifer, 246n3
Robertson, Craig, 253n66
Robeson, Eslanda Goode, 7
Robinson, Labada, 72
Rodgers, Lawrence, 40, 48, 248n17, 248n23, 248n26
Rollins, Judith, 97
romance, 45–46, 103
Roosevelt, President Franklin D., 229
Rose, Helen and Hugh, 134–35, 136
Ross, Marlon, 7
Royal Hawaiian Hotel, 167, 177, 180, 181, 185, 204, 233
Ruffing, Russell C., 265n10
Ryan, Harriet 264n4

Sabin, Will, 173–74
Salmon, D. E., 252n39
Salt Lake City, 60
Saltair, 59–60
Scheiber, Harry N. and Jane L., 208
Schoenherr, Steven, 252n35
Sedgwick, Ellery and: archive, 10, 174;

Faraway Women, 221; *The Happy Profession*, 198, 199, 225, 241, 262n38; Harrison, 51, 124, 164, 186–87, 188, 196, 198–99, 200, 221, 225–26, 264n30
Sellers, Jefferey M., 219, 264n16
Seville, 49, 65, 108, 110, 123, 135, 210, 214, 215, 216, 218
Shakespeare and Company Bookstore, 124
Sharma, Nitasha Tamar, 172
Sharpley-Whiting, T. Denean, 120, 139
Shellenbarger, Melanie, 250n74
Shepherd, R.S. and Ruth, 75, 252n52
Sherrard-Johnson, Cherene, 255n12
ship voyages, 99–101, 146, 158–59, 160
Sides, Josh 251nn12–13
Simmons, Beth A., 257n53
Sissle, Noble, 122
Skerrett, Joseph T., Jr., 170
Slater, Gene, 65
Smith, Ada, 120, 122
Smith, Ann Cunningham, 84, 130
Smith, Virginia Whatley, 243n6
Social Security, 228, 229–30, 245n6
Sojourner Truth Home, 68
South America, 211, 218–26
Soviet Union, 158
Spain, 214–18
Sparrows, 19–20
Spillers, Hortense, 12
Spingarn, Arthur B., 193, 240
Sri Lanka, 152
St. Louis World's Fair, 35
Stallings, Green B. and Lucy, 23
Stannard, David E., 166
Steedman, Carolyn, 10, 127
Stegner, Wallace, 59
Stein, Gertrude, 48, 192
Stein, Jordan Alexander, 244n29
Steiner, James, 183–84
Stepto, Robert, 49
Stermer, Martha, 261n29
Stowe, Harriet Beecher, 91, 238
Strong & Dickinson, 62, 63, 65, 229, 231
Strong, Frank, 63, 71, 72, 264n4
Sudduth, Charletta, 248n43
Sullivan, Noel, 192–93
sunbathing, 137–38

Sutton, Matthew Avery, 252n53
Sweden, 130

Tada family, 177–78, 187, 197, 205
Tada, George Yoshimato, 209
Tada, Kuni, 183, 208, 233
Tada, Lillian Shizuyo, 196, 198, 208, 209
Tada, Masakichi, 182–83
Taiwan, 159, 182
Taj Mahal, 74
Taj Mahal Hotel, 155, 258n25
Tamura, Eileen, 259n24
Taylor, Quintard, 249n52
Thaggert, Miriam, 53–54
Their Eyes Were Watching God (Hurston), 189
Thiebes, Raquel, 246n16
Thompson, Eloise Bibb, 68
Thompson, Era Bell, 6, 48, 56, 57, 120, 224
Todd, Bertha Boykin, 247nn53–54
Todd, Bertha Boykin, 254n10
Tomkins, Calvin, 256n42
Toomer, Jean, 46, 109
Toppozada, Adel, 150
Totten, Gary, 4, 243n8, 249n46, 255nn12–13
Towne, Charles Hanson, 192
Tragella, Umberto, 185–86, 204
train accident: in Alabama, 33–34, 110–11; in Czechoslovakia, 34, 111, 112
train travel, 52–54, 101, 153
Treaty of Dancing Rabbit Creek, 22, 24–25, 27
Trotter, Joe William, Jr., 248nn11–12
Trumbull, Robert, 260n41
Tubby, Roseanna, 246n23
Tufts, Mary Ann Dickinson, 73, 202–3, 204, 205
Twain, Mark, 154, 155, 169, 232
typing, 225, 240

Ulrich, Laurel Thatcher, 11
Uncle Tom's Cabin (Stowe), 16, 238–41
Uruguay, 222

vagrancy, 98, 102–3
Valley of the Temples Memorial Park, 233–34

Van Vechten, Carl, 48, 192
Van Wormer, Katherine, 248n43
Vasudevan, Ragunath, 258n28
veils, 151, 152
Velázquez, Diego, 217
Venice (California), 230–31
Villa Petit Peep (Waikiki tent), 177–79, 182, 185, 198–99, 200, 203
Volcano House, 168–70, 214

Waikiki. *See* Honolulu and Villa Petit Peep
Walker, Alice, 234–35
Wallis, Eileen, 56
Walsh, Owen, 64, 107, 110
Washington, Booker T., 115
Watson's Hotel, 13, 153–55, 169
Wells, Ida B., 54
Wells, Samuel J., 246nn23–25, 246n27
Welty, Eudora, 29
West, Dorothy, 4
Western privilege, 4, 142
Whitaker, Alma, 203
Whitfield, Eileen, 245n1

Wild, Mark, 66, 68
Wiles, Irving Ramsey, 125
Wiley, Nash, 29
Wilkerson, Isabel, 27
Williams, Brian, 244n20
Williams, Raymond, 117
Winks Lodge, 58
Wong, Mau, 182
world's fairs, 35, 144–45, 180, 237
Wright, Richard, 2, 42, 189

Yaeger, Patricia, xi, 21, 34, 101, 108
Yangon, 56, 152–53
Young, John Kevin, 10, 189
Young, Peter T., 260n38, 260n48
Young, Vershawn Ashanti, 109, 244n32, 245n36
Youngs, Tim, III, 158
Ypeij, Annelou, 264n17
YWCA, 54–57; Denver, 57–59; Honolulu, 168, 175–76; Los Angeles, 68; Kobe, 55; Madras, 156; Mumbai, 55–56; Nice, 129; Yangon, 56